Breakthrough In Cell-Defense

"Recently the secrets of glutathione (GSH) have been unlocked by researchers. Now comes the fascinating story of one of the major breakthroughs that will enable the public to benefit directly from these discoveries. BREAKTHROUGH IN CELL-DEFENSE has beautifully described what happens when you mix up hard work, serendipity and pure genius, the result ... a major medical discovery. Rarely have I read a scientific piece that reads as well as a novel! Thanks for bringing this story to light."

- Jimmy Gutman, M.D., FACEP, Emergentologist

"BREAKTHROUGH IN CELL-DEFENSE presents the fascinating story of the discovery of Immunocal® and establishes Dr. Gustavo Bounous as the father of trophic nutritional therapy."

- Thomas A. Kwyer, M.D., FACS Otorhinolaryngologist, Michigan

"Rarely do I go through a day in my office without being asked the question ... 'Doctor, what can I do to improve my Immune System?' Today more than ever, we crave for natural ways to enhance our health and boost our bodies'resilience. BREAKTHROUGH IN CELL-DEFENSE provides the answer by taking us on a fascinating journey through the defense system of our cells and into the life of Dr. Gustavo Bounous, one of the top Canadian medical researchers of all time. The discovery of a biologically active, natural, nutritional solution that enhances the immune system has unlocked the doors to one of the most exciting areas of medical research today. This has far-reaching implications for all of us in both health and disease. I look forward with anticipation to the future as we reap the benefits of this remarkable discovery."

- Dennis H.G. Forrester, M.A. Sc., M.D., CCFP
Family and Preventive Medicine

"BREAKTHROUGH IN CELL-DEFENSE provides an accurate reflection of medical research: the need for personal drive and perseverance, the combination of thoughtful, logical progression with serendipity, and the constant requirement for approval from regulating bodies and peers. This is told against the backdrop of a man, Gustavo Bounous, who was able to surmount his humble origins, work around societal and government restrictions, make and re-make his life in foreign lands, and succeed against popular conceptions. In many ways, the story of Gustavo Bounous is a true Canadian success story, a tale which has not been told often enough. The book is enlightening both to those who make the choice to take up the challenge of research, and to all of us who strive to improve the well-being of others. In addition, the book provides the reader with a comprehensible description of the role of glutathione in health and disease, and the therapeutic potential of Dr. Bounous' discoveries".

- Larry C. Lands, M.D., Ph.D., Director, Cystic Fibrosis Clinic Montreal Children's Hospital

"I've heard many times, 'If something sounds too good to be true, it probably is'. As a practicing physician and surgeon, I've found this to be usually true, but BREAKTHROUGH IN CELL-DEFENSE shows that Immunocal® is an exception. It has met and passed each test of science so far, and we're only at the beginning. Where this on-going glutathione research will lead us, no one knows, but it will be exciting and a great number of lives will benefit from it. We all owe a debt of gratitude to Dr. Somersall for bringing the life and work of a great man to our attention. We owe an even greater debt to Dr. Bounous, a gentle and kind spirit, physician and surgeon, and the ultimate scientist."

- Robert D. Biggers, M.D., FACS, Urologist, Colorado

PREAMBLE

Reading BREAKTHROUGH IN CELL-DEFENSE by Somersall and Bounous brought back a flood of exciting memories. In 1964-65 I had the extra-ordinary privilege of working in the McGill University Surgical Clinic with Fraser N. Gurd and Gustavo Bounous. It was the "academic enrichment" year of the McGill residency program in Surgery. I was working on a project related to cardiogenic shock alongside Gustavo Bounous who was studying small bowel changes following hemorrhagic shock. I immediately recognized that Gus was special in many ways. Watching him reason through the mysteries of the lesions in the small bowel was to see a master at work. He carefully dissected out this complex problem in a most innovative way, always challenging his own observations.

This book very accurately details the unravelling of this acute necrosis of the intestine which occurs in major burns and in all shock like states. More importantly Bounous took these observations one step further and pioneered the treatment of this lesion. His development of the elemental diet delivered via the gut was one of the important features in the current advances in surgical nutrition. Gus and Dr. Fraser Gurd would both be pleased with the central role thought to be played by the small bowel in the explanation of multiple organ failure following various low flow states. The important role of enteral nutrition in today's surgical intensive care units is a credit to the careful basic work of Bounous and Gurd.

Perhaps of more interest to me was the detailed description of Gustavo Bounous' early years. While I worked with Gus as a surgical resident and later as a surgical colleague I was not aware of his unique heritage. This clearly explains much of his behavior and his dedication to a career in surgical research. I now understand his feelings towards Louis Pasteur, Dante (which he often quoted to us as students in the laboratory) Galileo, Alexander Fleming, Beethoven and other personal heroes. Clearly his ability to overcome a variety of obstacles in life can be attributed to a character heavily influenced by his mother, and his childhood self-education.

Bounous' observations on enhancing the immune system followed very logically from his work on the elemental diet. His collaboration with Batist, Kongshavn and others to elucidate the mechanism of action of whey protein is a great example of the value of a critical mass of scientists in an institution. It also represents brilliant powers of observation, serendipity, dogged determination, and patient intuitive reasoning by Bounous. Collaboration with leaders in industry such as Beer has allowed further research and made the whey protein available in clinically useable form. The potential impact of this product has just begun to be examined but could revolutionize the care of the immuno-suppressed patient from any cause. Its role in cancer treatment, nutritional depletion, infectious diseases are all being explored. This approach utilizing a *nutritional intervention* rather than a 'drug' to which a resistance could evolve, is typical of Gus Bounous' perspective.

I am sure that Ada Bounous, H.B. Schumacher (Indiana), Fraser N. Gurd (McGill), and all of Gus' colleagues would be enormously proud of his current accomplishments and the recognition from the Medical Research Council of Canada. Gus in his own humble way says: "I did not discover anything. I was privileged to find a choice protein mixture, carefully derived from milk so that it remains undenatured and containing the critical precursor of glutathione."

David S. Mulder, M.D., FRCS(C)
Surgeon-in-Chief, MGH
H. Rocke Robertson Professor of Surgery, McGill U.

Other books by Dr. Allan C. Somersall:

Your Very Good Health: 101 Healthy Lifestyle Choices

A Passion For Living: The *Art* of Real Success

Your Evolution to YES!

Understanding The Evolution of YES!

Evolutionary Tales by Dr. YES!

Boost Your IMMUNE SYSTEM with the Latest

Breakthrough In Cell-Defense

An Amazing Health Discovery
for the 21st Century!

Dr. Allan C. Somersall, Ph.D., M.D.
with Dr. Gustavo Bounous, M.D., FRCS(C)

GOLDENeight Publishers
Atlanta * Toronto

Notice
Breakthrough In Cell-Defense is designed to increase your knowledge of a recent development in support of the immune system ... It makes a fascinating story.

Cover Design: Lightbourne Images

Printed by the University of Toronto Press, Toronto Canada

Library of Congress Catagloging-in-Publication Data: #98-073298

ISBN #1-890412-99-6

GOLDENeight Publishers
2778 Cumberland Blvd
Suite: 206
Smyrna, GA 30080

1-800-501-8516

Dedicated to

an angel mother,

Ada Bounous

DISCLAIMER STATEMENT

ACKNOWLEDGEMENTS

The authors are indebted to so many people for their valuable contributions that have made this book a reality: To the colleagues and reseachers whose knowledge and expertise contributed to the *breakthrough* over the past two decades; to Charlotte Hale for her diligent interviewing, researching and writing the preliminary draft which ensured the warmth and human touch to this story; to Carolina Loren for tireless efforts and long hours in preparation, while championing this project and following it through to completion; to Bea Hackett for the original idea; to my family (A.C.S.) for their patience, support and encouragement. Then there are those with whom we had valuable discussions and those who made constructive editorial comments. To one and all, we express our sincere thanks. We can only hope that we did your fine efforts and this noble scientific achievement the justice they deserve, so that it can have the healthy effect which many people in this modern world need.

TABLE OF CONTENTS

TABLE OF CONTENTS

breakthrough ▷ *noun* 1. **a decisive advance or discovery, especially in scientific research, opening the way to further developments.** 2. an act of breaking through something. 3. *psychol.* a sudden advance in a patient's treatment, usually after a long period without progress.

- Chambers 21st Century Dictionary

FOREWORD

The best way to find important facts, I tell myself, is to go directly to the source, if possible. Accordingly, I recently found myself driving from Toronto, Ontario where I live and practice medicine, along the Trans-Canada Highway to Montreal, Quebec. That day my every instinct told me that I was about to gain some of the most valuable information of my scientific life.

My appointment in Montreal was with Dr. Gustavo Bounous, Professor of Surgery and award-winning research scientist, who in 1993 concluded a distinguished career at McGill University and the Montreal General Hospital. Dr. Bounous, I found out, had fathered a major *breakthrough in cell-defense* which was still not common knowledge, even in the medical community. The popular media was not yet in its usual frenzy. But I knew a quiet revolution had been initiated. My curiosity was peaked.

Why am I interested in cell-defense?

I started my career in the field of chemistry which later brought me inadvertently into the vitamin industry. As I increased my knowledge of nutrition and food supplements and the relevant applications for health and disease, I came to the interface of nutrition and medicine, so I chose to go back to medical school. Later when I started my practice, it became very obvious that despite all the heralded technical advances of modern medicine and surgery, the real bottleneck to the health of industrialized societies was clearly in the area of health responsibility and prevention.

It has become clear in fact, that it is what patients choose to do for themselves between doctors' visits that is most important. Today the experts all agree. Central to all that one could do in the area of self-help to promote wellness, is the obvious need to boost the immune system by healthy lifestyle habits. That would in general, enhance the body's capacity to defend itself and at the cellular level, nothing could be more effective.

It is fundamental, if not apparent, that *cell-defense is the best form of self-defense.* Therefore, to find a way to improve universal cell-defense must obviously have dramatic implications.

So learning of this new breakthrough by Dr. Bounous and his co-workers, sparked my keen interest. I wanted to know the details.

As I settled into the pleasant drive, by car, like just one cell in the fast-moving body of traffic, my mind sped even faster, criss-crossing numerous avenues of speculation. What sort of man would Dr. Bounous be? What would it be like to interview this giant of medicine, the recognized author of several astounding findings? For that matter, I mused, what might it have been like to interview Dr. Jonas Salk, Dr. Louis Pasteur or Dr. Alexander Fleming?

Why would I dare to place this Dr. Bounous among such historic company?

Good question.

Quite simply, Dr. Bounous had discovered a facile way to boost the human immune system by means of a unique bovine whey protein concentrate that was destined to make astonishing medical news. He and his colleagues had spent almost two decades doing fundamental research which led to the development of a proprietary process to isolate some critical undenatured proteins. Laboratory tests had produced convincing evidence that this particular whey product dramatically enhanced the immune system via cysteine links to glutathione. That is *the* critical agent of cellular defense against oxidation, viruses, toxins and other marauders. Another of nature's secrets had been unveiled, it was simple yet profound.

I knew from the published literature that the select whey nutrients had also helped to stave off select bacterial infections and even arrested the course of chemically-induced carcinomas in laboratory animals. Further tests showed that the natural life span of mice had actually been increased by up to one third. And that was just the beginning.

The implications boggled my imagination. What about AIDS and human cancer? ... I wondered ... Or hepatitis B? Or the various failures of the aging human body? And what might this immune system build-up mean for such diseases as chronic fatigue and environmental illness, allergies and the common cold, or even breast, colon and prostate cancer? The unique proteins optimized intracellular glutathione which is a proven super anti-oxidant and more. What else could that mean? Even widespread cardiovascular disease and many degenerative conditions have premature oxidation steps implied in their pathophysiology. Would something that could controllably modulate glutathione therefore have any widespread role? I had a thousand similar questions for Dr. Bounous ...

Why ...?

Soon I pulled up at the offices of Immunotec Research Ltd. That's the successor to the company that first developed and produced the select whey protein concentrate, patented it worldwide with staggering method-of-use claims, and now markets and distributes it. This same milk whey product that had generated such dramatic results in the laboratory - is now available for human use. Dr. Bounous gladly serves as the Director of Research and Development for the new company.

As I walked in that day, I could not have imagined that the quiet man seated behind the paper-strewn desk in that small, plain, white-walled office, was soon to enlarge my thinking and literally redirect my life. The scientist instantly impressed me as a warm, unhurried, disarmingly unassuming human being. A sweep of gray hair above a broad brow accentuated his dark brown, unusually expressive eyes. He was alert and very engaging. Approaching age seventy, the lean, active Dr. Bounous still looks ahead to some of his most important work.

That work now entails overseeing continuing research projects, while putting in long office hours, traveling thousands of miles, and answering urgent questions posed by other physicians like myself. The doctor shows no signs of slowing down; he is still very excited by uncovering a vital link to cellular health, his golden key to improving the human immune system. He is proud of his achievement, yet he remains modest, too modest to shout.

But he does have something to shout about. If he did not choose to tell the story, who would? That was the obvious question. Yes, who would chronicle the serendipitous events that led to this breakthrough and who would dare to point forward to its promising horizons?

Why not document the breakthrough?

After that day in Montreal, I wanted to write a book with Dr. Bounous which would explain in detail not only the scientific facts but also the profound sgnificance of the Bounous discovery. The doctor graciously consented. On subsequent visits, however, I revised the plan. I came to realize that there was far more to the story than glutathione, lymphocytes and biochemical analysis. The human elements were equally electrifying.

The story of the Bounous scientific breakthrough includes one man's personal journey through curiosity, frustration, disappointment, apparent failure and ultimate success. It involves dogged work, loneliness, luck, incredible 'accidental' findings, and an exceptionally distinguished group of colleagues who labored alongside the scientist in a remarkable display of teamwork and loyalty.

The story within The Story is as compelling as anything Hollywood could produce. At times, indeed, the reader will find that ***Breakthrough in Cell-Defense*** reads more like a suspense novel and moves faster than a *Movie of the Week*. It comes to a climax and humanity wins. Most importantly, however, the story behind this far-reaching discovery rings with scientific truth. It sets a new standard.

I offer this real-life story of Dr. Gustavo Bounous and his landmark contribution to human well-being, with the hope that it may touch your life as deeply as it has influenced mine.

Allan C. Somersall, Ph.D., M.D.
Toronto and Atlanta

1928

"... the handwriting ..."

1.

MAN AND DESTINY

Perhaps the only exceptional thing about the birth of tiny Gustavo Bounous in Luserna, Italy on July 10, 1928, was his parents' boundless joy. Sure it was a boy! The screaming infant born to Carlo and Ada Bounous on that humid, Piedmont-splendid, midsummer day was to usher in a bright new hope – not only for his parents, but for the entire Bounous tribe. So they hoped. They prayed.

"My sister Luciana had been born two years earlier, and my sole uncle had all girls, so our family placed their hopes for our collective futures on me."

With those words, Dr. Bounous lounged back, looked up to stare at the ceiling and then risked a soft smile. After all, that was before the days of woman's liberation and only men could be expected to shoulder the huge responsibilities of lifting a family up and providing meaningful security.

I could see that Dr. Bounous became much more animated and passionate as he reflected in a moment of true nostalgia.

"My mother, in particular, always said I must become a doctor. What more successful career could there be?"

In those days, it meant more than wealth and prestige. It was a symbol of accomplishment through sacrifice. Being a doctor was a life of service with pure devotion and hard work. But it was indeed to arrive: to become somebody and to command respect from everybody. A family could fly on those professional wings. His mother placed that thought in his head very early, and said words like those very often.

As fate would have it though, the innocent object of the Bounous family's hopes happened to be born on the cusp of chaos. Less than a year later, the Great Depression would smite the United States following the remarkable stock market crash that thrust a great nation into poverty. The economic ripple effect quickly extended to Europe, where the population was still recovering from the devastation of World War I. No surprise then, the depression hit Italy hard, and times got rough again.

Before little Gustavo could even learn to walk, his country had already begun to totter. A calculating politician named Benito Mussolini seized advantage of the economic and social turbulence in Italy, at that time a democracy, and introduced Fascism as his answer to the unrest and chaos. Mussolini was shrewd and charismatic. He succeeded in his own design and the rest we know as history: Italy became Europe's first post-war totalitarian nation, and later led the way into the multiplied horrors of the nearly cataclysmic World War II.

"Tito," Ada Bounous would tell her teenage son again and again during those first dark years, "study hard to prepare yourself. When times get better, you must be prepared to pursue medicine."

What other hope could she offer? She had seen careers from both sides. The Bounous family had their fair share of education – Gustavo's father was an engineer, his uncle Guido a businessman. But during Gustavo's early years, even the middle class was feeling the tremors of economic and social insecurity. Education itself was no guarantee of survival, much less of success. But Ada hoped – she knew - a doctor's life and livelihood would be different, in good times and in bad. So she dreamed for her only son, and she gladly made her dreams known.

"My father was anti-Fascist, so he could find very little work," Dr. Bounous explained. "Occasionally he would get a break to teach mathematics, but otherwise he did very little with his life. He lived and died a quiet man."

And just as well. Mussolini was sending gangs of armed thugs to assault dissenters in the streets. He was hounding and beating working men and even women who dared to oppose his views. No doubt, Ada urged her husband to keep a low profile. And that he did.

As the boy Gustavo grew, he observed his bright, strong and well-educated father slowly fade before his eyes. His dad became someone who believed less and less in his native country and then in his own personal significance and talents. He capitulated to terror with indifference. Meanwhile his Ada, a slender, dark-haired young beauty with a cameo-like profile, seemed to increase in tensile strength. Drawing her two children about her like a hen, she insisted they must live and practice 'hoping against hope.' Her strong spirit and personality took root in her impressionable younger child, her 'Tito.' Year by year he grew more and more into the role of an uncrowned head of the family, willing to take on increasing responsibilities and independence, always desiring to help his mother.

"At first I did not know how bad things were," he remembers. "I spent time playing in the large garden outside our small house. I enjoyed the outdoors. I had a little bicycle and I would play soccer by myself. In fact, I think I did most of the things that small boys in our village did. I had fun and after all, I had my family. Our parents tried to hide their worries from my sister and me, but with war and violence encroaching, of course that could not last forever."

It was during these early days that a family friend started to call the boy "Gustavito," a diminutive of his name which, conveniently it turned out, became shortened to "Tito." That affectionate name stuck, but Gustavo gave it a different meaning in a different setting. He was so gentle and generous that his pet name could become a symbol of contrasts.

Dr. Bounous at times during our conversation, would characteristically lower his voice. Yet he spoke freely and emotionally. He had vivid recollections.

"Growing up in Luserna, in Italy's Piedmont, was quite delightful. We were in a peaceful valley, lying low and verdant amidst the hill country. I remember the sounds of the village, like the angelus bells echoing through the mountain air and the rhythmic sounds of the local train.

"My father somehow never really succeeded - understandably so, since he was always never fitting in. As a consequence, others in our family helped to support us. But those hardships made my mother even more determined that I should become a doctor."

Luserna was itself a real picture of living contrast. Imagine some of the most lavishly beautiful foothills and mountain ranges of Europe. Noble architecture lined the ancient streets, where some of the finest churches and shrines of Italy were on display. Yet the presence of Fascist troops in black shirts and heavy boots overshadowed this pure serenity with the violence of war.

"The pain began for me in 1940, when I was only twelve," Bounous recalls with a shudder. "We moved from the city of Torino, where we lived by then, to a little village called San Giovanni. I still remember it so well. It was engulfed by those beautiful mountains, the Italian Alps. My mother would climb some steep paths on the hillsides to get away and to gaze at the magnificent views, alone with her thoughts."

God only knows how this matriarch philosopher rationalized her circumstances. What made the wheels of fate turn? Where was her family heading anyway? She would muse long and hard but she always managed to keep a smile. She was a survivor at heart.

Tito himself seldom climbed. No one really knew then, but from his early childhood, he experienced a minor but troublesome hip problem. It was probably a congenital misalignment of both ball and socket joints and it made him dislike making such arduous physical efforts. This chronic condition would later impact his career but as we shall see, it was a blessing in disguise. The mature Dr. Bounous would then see beyond his limitation to reach beyond his grasp. Tito, the young child, thought much. He hardly roamed the hills, but still like his mother, he felt the lure of distant horizons.

"A two-hundred-year-old chestnut tree grew just outside my bedroom window. I remember it well." Dr. Bounous's eyes would open wide and sparkle with these words. "From its branches I could see in full view, a range of mountains which I learnt quite early, was on the border with France. Beyond France, I knew, lay a great ocean, and somewhere out there beyond that ocean was America. America! As a boy, I was free to dream. And that I did.

"Someday I would go to America."

Does the man make the times, or do the times make the man? Hemmed in by wartime constraints, Gustavo Bounous, like his mother, turned inward and reveled in his own dreams and schemes. He ruminated over his ideas and ideals. One would presume that a young boy could not prevail against the indiscriminate forces of war. Nevertheless, in the right circumstances, even the most powerful armies cannot overcome the power of a determined boy's dream. Tito was determined to achieve something; he knew not what.

Ada's Tito was stuck in a little isolated mountain village where there was little apparent opportunity. But he began building his own inner fortresses. The boy read voraciously, devouring the classics. There, in that breathtakingly splendid village of lofty mountain heights and dazzling beauty, yet that uneventful place which offered no chance for personal ambitions to flourish, he was isolated, cut off from most of the world's affairs. It was in this place which many would deem hopeless, that from a young man's choice of reading – and hence, the direction in which his thoughts would go – a man and his destiny had begun to take form.

More than half a century later, Dr. Bounous describes with affection the medieval Roman Catholic churches near the village, and the ancient roadside shrines that date back to the First Century. He still could picture those meandering footpaths his mother walked, which also bore unseen footprints of some of Christianity's earliest pilgrims and martyrs. Gustavo to this day is charmed by history, the total panorama of the human drama.

The Bounous family itself, however, was Protestant. The Waldensian sect was a small group of pilgrims who followed the strict teachings of Christ and eschewed the ritualistic Roman Catholic church. They had for centuries – far earlier than Martin Luther's Reformation movement – disengaged from typical practices of that church. For centuries, those earliest Protestants had been hunted down like animals. They had been beaten and killed. They had taken refuge in the French, Italian and Swiss Alps, hounded from place to place and reviled as heretics.

Gustavo Bounous had been born into that simple faith. Indeed, his great-grandfather for years headed the church, which even today numbers some fifty thousand souls. The boy who surveyed the world from his chestnut tree could see the very hills where his pious forefathers had died for their faith. He had so many questions and so few answers. More tangible, though, and much more significant to Gustavo, was the collection of letters his great-grandfather had written back to his family and church while on three separate fund-raising trips to America. He was enthralled by them.

"My great-grandfather found much sympathy in America for those who were persecuted because of their religious beliefs, and he wrote that Americans were very good and generous people," Dr. Bounous observed. He felt a kindred spirit there. It's easy to see why these letters aroused a young boy's curiosity, hope and sense of adventure. And it's even easier to imagine that when young Tito became fascinated by America's most favored president, Abraham Lincoln, he found someone who he thought was very much like himself.

Abe Lincoln was born in a now famous Illinois log cabin in 1809. He lost his mother before the age of ten and his sister before he was twenty. His first venture in business was a general store partnership which failed within a year. He was first elected to the State House at age twenty-five and began to study law. He fell into a deep depression. He failed in his bid to become the Speaker and four years later, got married to his sweetheart only after breaking the initial engagement and enduring another depression before reconciliation.

The unattractive Lincoln failed to gain nomination for Congress the following year. He set up his own law practice with a partner but not long after, he lost his second son at only age four and his father the following year. Five years later he was back in politics but lost two different bids for the Senate. Finally he triumphed in becoming the sixteenth President of the United States.

He grew in stature and wisdom, from the illiteracy of his own parents to the immortalized Gettysburg address. He migrated from the lonely log cabin to the White House, but only through a compounding series of failures and adversity. He became the prince among American presidents, one who saved the Union. The ordinary became extraordinary.

Lincoln, of course, had endured the same sorts of hardships and privation that young Bounous knew. The boy Abe had walked miles on foot to borrow books which he had to read by firelight. He was ridiculously poor … and honorable. Lincoln read everything he could lay his hands on. He thereby taught himself. The same Lincoln overcame his undue share of disappointment and loss. And after rising from such hopelessness to become President, he would later write …

"*All that I am, and all that I can ever hope to be, I owe to my angel mother.*"

That hit home forcibly to Gustavo. The events of their very early lives were so common. War or no war, Gustavo Bounous decided that someday, God willing, he would cross the imposing mountain range, then cross France, and finally cross the Atlantic Ocean and get to America, the land of Lincoln.

Today he reflects with maturity. He discerns the roots of his own personal philosophy.

"My biggest reason for coming to America was to escape human ignorance. I had endured in my young life both fascism and war and these were to my mind the epitome of that ignorance. Ignorance is the ultimate curse of humanity. Only in the pursuit of real truth can we find freedom and redemption."

1928. Of course Dr. Bounous, like the rest of us, cannot remember the year of his birth. It so happened that ironically, there was then another and more important beginning. In fact, that same year also ushered in a momentous medical event. It was an accidental discovery, to be sure, but a world-changing one – Dr. Alexander Fleming's discovery of penicillin.

At. St. Mary's Hospital in London where he had a germ lab, Fleming on one occasion carelessly left his laboratory window open and his petri dish cultures exposed, while he went out of town. When he returned, he noticed that in one of the dishes where he was trying to grow staph germ cultures, there was a "contaminating colony." Amazingly, the mold colony clearly had killed the deadly staph germs. *An ordinary organism had wiped out deadly bacteria.*

In 1928 no one took much note of this happenstance, one which would have life-saving and world-changing consequences. And ironically, a dozen years later, as war rolled like a death tide across cities, nations, continents and almost around the world, Dr. Fleming's still new medical discovery - one destined to save countless millions of future lives - still waited in history's wings.

In 1928 moreover, nobody could possibly foresee or imagine that two decades later, the paths of this pioneer English doctor in London and this newborn Italian infant in Luserna were destined to cross. But that event, like the handwriting of providence, would hold great significance for Gustavo Bounous. It would be more than a meeting of the minds. It made for inspiration across a generation and set the stage for another surprising breakthrough …

1938

“ … against all odds”

2.

GREAT IDEALS vs. BOUNDLESS IMPOSSIBILITIES

That fateful year, 1938, the prelude to a war estimated to have cost the lives of some thirty to forty million people, both military and civilian, nearly cost the world its soul.

War is senseless at any time and usually devoid of purpose since it generally settles little or nothing. The conflict of egos and stubborn wills only leads to the carnage of young and innocent life. The parochialism and folly of warring factions create only pain and suffering everywhere.

Ada and Carlo Bounous experienced the war firsthand while raising their young family. They saw enough … too much. They could get angry but they could never get even.

"Tito, you must become a surgeon," Ada decided. "Learn to save life and not destroy it. You will help relieve the pain and suffering of mankind and so weep with those who weep."

The worse the world around them became, the more strength this heroic woman seemed to exhibit. She was hardworking and always positive. She devoted her thoughts and her sometimes staggeringly bold ideas to her family – and especially to her son.

It was Ada who continually reminded young Gustavo, her gangling teenage youngster with the sad eyes and broadening shoulders, that the war surely must end soon. How much longer could Europe endure such devastation?

"We must pray for peace. But we must prepare for after the peace. The moment the guns cease to fire and no more bombs explode," she emphasized, "our family like everybody else, must struggle to regain our economic security. We must restore our dignity and bring our lives back to some kind of normalcy."

The time to prepare was *now*. For their family, a lot would depend upon Tito's ability to succeed. That's how Ada framed her world. Young as he was, Gustavo felt those responsibilities keenly. He was growing and the daunting challenge only accelerated his maturity.

By now his little family of four had for some time become part of the larger Bounous enclave in San Giovanni. They lived in one of the four family houses surrounded by small garden plots, a few animals and a bevy of girl cousins. Tito and his sister Luciana grew up as the poor relations, those who always needed some kind of help. Their good-hearted relatives never spoke of that, of course, but such frequent humiliation stung the sensitive young man. Gustavo could not help but contrast his father's bitter defeatism and resignation with his mother's valiant efforts to support her children. She wanted the best for them and kept looking to the future in hope. Even against hope, she believed in hope.

During those years, as wartides ebbed and surged around different parts of Europe, the fortunes of people in Italy's lovely Piedmont bobbed like corks on their country's political tides. Within the brief years of Tito's childhood, his country declined from a promising democracy, into the hands of an opportunistic *duce* who plunged the nation unwillingly into fascism, dictatorship, totalitarianism, and – the ultimate humiliation – an alliance with Nazi Germany.

"One day the Fascists came and took my mother away," Gustavo Bounous still relates with compelling emotion. "They accused her of working for the enemy! That was enough … if only for them. There was no evidence, no trial, no defense … no justice. Justice has no place in war. War has no feeling. It is all brutal and cold."

Dr. Bounous is at a loss for words. He clears his throat and moistens the eyes as they speak volumes to my listening heart. After all, that was his mother.

The fact was, Ada Bounous was most caring and warm. She did anything her hands could find to do to help feed them all. She was hardworking and conscientious about everything she did. After Ada managed to convince her accusers that her motives were economic, not political, they returned her to her terrified teenagers.

"We lived in constant fear," Bounous said. "Not only did we fear the Germans, but also we dreaded attacks by the odd gang among our own countrymen. Not all of them fought for freedom, some even at night swept into homes to bully and plunder at will."

Within this miasma of fear, deprivation and the awful sense of loss of national honor, Gustavo Bounous, the skinny boy with the improbable call on his life to become a surgeon, spent his formative years. In at least one way, however, the family was fortunate. There in the foothills and mountains of one of the most gloriously beautiful areas of Italy – some would say, of Europe – there existed for most of the time, at least some small measure of peace and sanity. There was a tranquility in nature, immune to the vicious turbulence in society at that time.

From today's perspective, one marvels at the fortitude of women like Ada Bounous. Day by day she watched her husband become a silent casualty of war. He had been wounded in the war, true, but he had recovered. His unseen wounds, those of disillusionment and hopelessness, never seemed to heal. Thus the war cost Ada her husband as surely as if she had lost him to a bullet or a bayonet through his heart. When a man's spirit dies ... he dies, if only on the inside.

But the Bounous nuclear family survived. Ada was a fortress, a rock of Gibraltar. Young Gustavo and his sister found their solace there.

No wonder Ada Bounous seemed larger than life . Amidst the changing scenes, she still seized precious bits of time to climb the rocky, twisting footpaths for a few precious moments alone. In her loneliness, like Joan of Arc she found her strength. Perhaps she merely thought about the right time to plant her *flageolets* or *pomodori.* More likely, she planned and dreamed for hours on end about her children's futures. Certainly, however, she returned from those climbs exhilarated, with new hope and determination. She would pass both on to her children.

The San Giovanni village provided few opportunities for jobs or even much education for a boy like Tito. Never mind though, there were those educated adults in the family, and great books to read. Fortunately, the boy liked to read. He devoured the library. He had an excellent, thoughtful mind, a questing intellect. Added to this appetite for knowledge was a loving mother who never failed to exhort and encourage him.

'Somehow', Ada dreamed, *'Tito must become a surgeon. After all, was there not some proverbial distant cousin who was a professor of surgery? Surely something must be possible ...'*

Gustavo buried himself in the books. He loved to read about America and adopted Thomas Jefferson and Abraham Lincoln as heroes. He read his great-grandfather's letters, over and over. The yellowed pages, by then nearly a century old, had a charming influence. They connected. He came to admire such a commitment in his favorite ancestor who risked his life to head the Waldensians. Those dissident people fought off the cruelties of the powerful Vatican which periodically issued orders to flush them out and exterminate them. They refused to be crushed.

If the reality of war became a fact of life for young Bounous, so had it been for his pious ancestors too ... not only his great-grandfather, but at least five centuries of others before him. Bounous was feeling a part of history. He was proud of that heritage. Yet he kept thinking of America. Obviously America's Lincoln and Jefferson also had risen above war ... yet still achieved greatness. There were mentors he could emulate, at least in principle.

Though the political fortunes of nations did not stir the adolescent Gustavo, humanism and heroism did. He was fascinated by great biographies. His reading and thinking ranged far beyond any national boundaries and as he read, his mind gravitated toward people of character and accomplishment. His favorite composer of music, for example, was that great German, Ludwig van Beethoven. His music transcended his origins. Not only did Bounous recognize Beethoven's musical composition as those of a genius, but the man himself was a titan.

Bounous's eyes gleam and his face becomes animated as he describes how Beethoven worked against his ever-encroaching deafness.

"Just imagine!" he exclaimed. "He could not even hear the music he composed, except in his head." Perhaps that made him attuned to sounds more celestial. They went beyond the mere cerebral.

"When he conducted his Ninth Symphony for the first time, the man was almost completely deaf. Imagine, the most magnificent symphony ever created ... and afterwards the great composer turned around only by custom, to face the cheering, applauding audience. He actually could not hear the spontaneous outpouring of adulation."

Ask Gustavo Bounous today to describe his personal heroes, and all bear the same characteristics: genius, creativity, humanity, idealism and so on, as you'd expect ... and in each life, the necessity *and the ability* to prevail against seemingly impossible odds.

What role did this boyhood idealism eventually play in leading young Tito into a distinguished, even brilliant, scientific career? Two Italian "heroes", possibly the most impressive of the lot, provide strong clues.

"Who?" I wondered out loud.

"Dante", Bounous says, almost reverently, "and Galileo."

He leans back once more in his chair, closes his eyes and begins to quote stanzas of "*The Divine Comedy*" in mellifluous Italian, then translates into English so the hearer can understand. He revels in this epic poem which represents a journey of the human soul. It details a narrative of a journey through Hell, up the mountain of Purgatory, and through the heavens into the presence of God. This is no "*Pilgrim's Progress.*" It is transcendent and inspiring, and Bounous is propelled into mental flight.

This gigantic work, which the young man had read again and again, and now can quote from memory decades later, contains much that Dante Alighieri knew of theology and philosophy, of astronomy and cosmography, and interesting tid-bits from numerous other branches of learning. Dante became the philosopher – king of a young Bounous' world.

As for that renaissance man, Galileo Galilei, considered by many as one of the most brilliant scientific minds of all recorded in history, he represented that individualism that gloried in emancipation. The great individual always stood out from others. He shaped his own destiny. He had *virtu*, the quality of being a man (*vir*, "man"), in the sense of demonstrating uniquely human powers against ignorance.

Galileo accepted the world as it was, but was always learning broadly and excelling in all that he did. Ironically, Bounous recognized and admired that same spirit of free enterprise and rugged determination which motivated the early American pioneers and later built the envious land of his dreams into a great empire that he would long to see. Perhaps in America he too could find freedom.

Dante and Galileo ...visionaries and thinkers like those were Tito's real teachers; men long gone, yet immortal; men who attempted to see life as a whole, trying to grasp its enormity, reverencing its mysteries, while diligently studying to understand its complexities.

As Tito matured from his late teens into young adulthood, he was intoxicated by great literature and the vast world of ideas. The aspiring Gustavo Bounous longed to explore these new horizons of the mind. He was drawn to literature ... perhaps to become a professor of literature, a lecturer or a critic ...

Despite this secret yearning, Gustavo Bounous never opposed his mother's deep-seated dream for her son. His was an attitude of deference as much as resignation.

Gustavo's learned hero Dante, after all, had embraced all of the literature, philosophy, science and religion of the Middle Ages. His intellectual diet was eclectic. Gustavo realized that to begin one's lifetime of learning with the study of medicine, then, surely was to pursue as great a discipline as literature or any of the others. Learning is learning. It is all important.

At least, he had choice. Ironically so.

The war ended in 1945. Gustavo Bounous, in so many ways a product of that war, recalls the day when, for him the peril was over.

"They drove up in Jeeps - the Americans," he said. "Out of the Jeeps jumped those G.I.s in their olive-brown uniforms. They looked like military police ... very tall. I was seventeen, and terribly excited. I will never forget it. Those soldiers – some white men, some black – jumped out of the Jeep and spoke to the crowds. 'It's all over, boys!', they shouted.

"Those were the first Americans I ever saw, and the first English words I ever heard."

Fortuitously (or by divine appointment, who can tell) - but in any case - fortunately, the son of Carlo and Ada Bounous, that renaissance man was still not yet eighteen, having been born in 1928. That was a narrow escape. Had he entered the world even one year earlier, there likely would have been no Dante or

Galileo, no Beethoven or Lincoln for him. Instead, Gustavo would have had to become a soldier. And who knows what that might have meant. By just months, he missed military conscription.

Rather, Gustavo Bounous gained academic freedom to become a man of destiny.

1948

" … to feed on hope … "

- Spencer

3.

EDUCATION OF A DOCTOR

"Thank the good Lord, he's on his way now."

Once eighteen-year-old Gustavo had been accepted into the University of Turin, Ada Bounous at last could breathe that deep sigh of relief. This mother had involved herself almost sacrificially in her son's development and education. Indeed, at her urging, he attended school one summer and advanced himself by a full year. Then the dream moved that much closer. Now he was established in a venerable and well-respected university. His possible future in medicine had at last been launched.

"It was easy to get into the university but almost impossible to get a job following graduation."

That's what Dr. Bounous remembers of matriculation but that was far from the whole story. Strapped by the enormous burden of waging history's most costly war, Europe, North America and Asia alike faced the burdensome tasks of rebuilding not only individual lives, but whole shattered cities and entire economies. Most institutions were in upheaval and some were in ruins. Nothing could be taken for granted. Nothing.

Gustavo Bounous entered university life at the precise moment in history when millions of former military men were flooding back into the halls of academia. Not only were universities everywhere bulging at the seams, but once graduated, the war-weary students would eventually compete for the few new positions which would be found in still-floundering job markets.

If the outlook appeared bleak, young student Bounous did not complain. His heroes would not approve. He enjoyed his classes and found science and medicine absorbing. Well used to academic disciplines, his courses seemed easy enough to master and interesting enough to merit his hard work. Through medicine, Gustavo at last allowed himself to hope, perhaps he really *could* build a better life for himself and his family.

No doubt many of his colleagues carried the same dreams and zeal into university life. The pensive, analytical Gustavo however, almost immediately came face-to-face with some disillusioning realities of college life. Whereas his idealism flourished, he got an awakening. Italy's medical doctors whom he began to know seemed to represent an elite and most often, an almost unbelievably privileged class. Bounous observed that lowly doctors-in-training fell into two distinctly different categories: there were some few from wealthy and influential families, and then there were others, like him, whose families had no heritage of prestige and essentially, no name. Indeed, at that time Gustavo's father was jobless.

He extrapolated his observations to the establishment. The Italian establishment. The Medical establishment.

Early on, Gustavo learned that Italy's medical community was ruled by those he called the "barons of medicine." These were the chiefs who recommended young doctors for jobs, not on merit, he insists, but generally according to one's family, their social or financial position. In his own case, he had no asset of that kind going for him. He could labor and learn, even distinguish himself academically, but he lacked the worldly connections that could make those labors pay off.

"The job you got depended on who you were and whom you knew," Bounous reflected. "It was all political."

Did this create fear in him? Would he retreat to farming in the countryside?

"Not really. I saw they were training many hundreds of doctors who could have no place to go, no certain hospital staff privileges. Even young *doctors* would have to struggle. They might become mere human technicians, enslaved to long hours of demanding grind with little recognition and inadequate reward.

"Nevertheless, I was in a good university, and I continually hoped that something good would happen. Day after day I studied and worked. I thought of my 'heroes' – Beethoven, for example – and the greatness of such men, whose lives reflected a triumph of mind and spirit over material circumstances. Hardship, after all, teaches you to be humble but also resilient."

The cultures of Renaissance Italy and post-World War II Italy seemed to collide within the slender, sad-faced youth. On the one hand, his reading, thinking and early idealism had created in him an enthusiasm for an open society of pure democracy – with opportunity for all, and promotion according to one's merits. He chose to celebrate excellence.

"Reality was, however, that many doctors I came to know – sometimes even eminent physicians – often seemed to strut around, seeking to promote themselves and to continually jockey for positions. They were ego-tripping. Such pretense made me disgusted. They often dressed up in fancy clothes and were driven around in limousines. Why? For what purpose? I thought doctors would be caring and compassionate servants who cherished their privileged vocation. I expected them to be genuinely curious about the human condition, reverent of life and humbled by their own inadequacy in the face of death. But I found too many doctors who acted pompous and grandiose ... ridiculous."

Such observations cause Dr. Bounous to be beside himself. At times he can hardly contain himself. He still wrestles with reality as he finds it. He gropes for light on his common path.

"There was one scientist I remember well, a Jewish man, who had been kicked off the university faculty during the war because he was a Jew, but later he was reinstated. From my perspective, if ever there was a really distinguished doctor, Giuseppe Levi was one. He was proficient in anatomy, embryology and histology. He was truly intelligent and very kind. I could perceive this man's greatness. There was no pomposity or arrogance in him, only humility and sincerity. He was not like those many other doctors I came to despise. I wished there were more like him."

University then, for the aspiring doctor Bounous became a clearing house for character. It was a filter through which he could sift the qualities of human greatness he believed one must have and could achieve. Very early he had decided that mere wealth and show were not worth pursuing. On the other hand, real greatness, he concluded, resided in those who accepted their circumstances in life with courage and resilience, and taught themselves through perseverance to triumph over personal adversity.

The two roads of life diverged. He would take the latter path.

At age twenty Gustavo Bounous, lean, sad-eyed, usually lacking even enough lira for a simple meal or a bus ride across town, had determined his personal value system. He sought no limousine, no chauffeur, no rich apparel, no sumptuous meals, and absolutely no position or prestige. In any case he realized that there was little likelihood that any of these things would ever come to him.

"I was a nobody." Dr. Bounous declares, resonating with nostalgia. "But I never felt embarrassed by that fact. I was content with my life. Nevertheless, it always seemed that somehow, things might change for the better. I knew not how or why.

"Day by day, I fed myself with that hope. It was the leftover from what my mother had fed me before."

One day during Gustavo's first or second year at university, things did change. The entire school, and indeed the entire city of Turin, was agog at the news that a huge Medical Symposium was to take place there, the first such event since the war.

To add to the importance and honor of the occasion, a famous scientist would be celebrated. Dr. Alexander Fleming, discoverer of the "miracle drug" penicillin, was to be the main speaker ... an occasion which made all Europe it seemed, look to Turin with respect.

Gustavo determined that he must attend this meeting. Never mind that he was nearly penniless, or that he was only a university student and a newcomer, at that. Here he could come close to excellence and perhaps to live greatness.

The meeting hall was across town, the place would be packed; there was no one to sponsor him or help him get in. Still, the would-be scientist plotted and planned to attend. He had already determined that an opportunity this important would not escape him.

It took a *bravura* effort, including two bus transfers across town, and a *bravura* performance, during which he nonchalantly entered the hall, *sans* credentials, camouflaged among a group of doctors. Thus young Bounous found himself at a meeting many doctors would have given their stethoscopes to attend.

The occasion profoundly influenced his life. The program proved electrifying on more than one count.

"Doctors from everywhere filled the stage and faced the audience," Bounous recalled. "Most of them arrived by chauffeur-driven limousines, but this time not dressed in their finery. For this symposium, these important physicians, the magnificent doctors all Italy was supposed to revere, actually appeared on the platform in white coats. They were making a statement.

"This seemed incredible to me, their posturing and pretense. They were attempting to flaunt their credentials even though they were hosting a celebrity who stood now head and shoulders above them all. It proved nothing.

"At last came the main speaker, Dr. Fleming. When his name was announced, the auditorium became totally quiet in anticipation of the great man. All Italy seemed to wait with rapt attention."

Excited, young Gustavo wondered what to expect of the legendary scientist whose fabulous discovery had saved the lives of so many thousands of Allied soldiers' during the recently ended, long and bloody war. What would he look like? Would he be a powerful speaker, with confidence and wit? What persona would he project? He held his breath and then ... Sir Alexander Fleming stood before them - a slender, sandy-haired Englishman, neither handsome nor impressive of stature. He seemed at ease and very real. As he spoke, this youthful listener marveled at the scientist's detachment, understatement, and lack of credit-taking during the astonishingly low-key account of perhaps the greatest medical discovery of the Twentieth Century. It was greatness personified.

Bounous was electrified.

The drama of penicillin, Bounous knew, was largely played out on the American stage to a worldwide audience during the early days of World War II and following. Though the momentous discovery actually happened a dozen years earlier in England, Dr. Fleming at the time essentially deemed it novel and exciting – a common mold organism that could overtake and obliterate dangerous, even deadly bacteria! – but virtually impossible to utilize.

Serendipity ... Irony ... Blind Folly.

A major breakthrough went unnoticed for years. It was too simple, too common-place, too ordinary.

How could this discovery evolve into a viable pharmaceutical product? How could it be transformed into something useful, measurable and most importantly, *safe*? Who would fund the necessary expensive research? And how many millions of dollars might it cost to make any commercial application happen, if indeed it could happen at all?

Above all, was it really likely that an untidy bacteriologist's laboratory accident could ultimately change the course of modern medicine? Far-fetched in the extreme!

Few at first could follow the real-life cops-and-robbers saga of bringing penicillin into everyday use. The twists and turns in this incredibly expensive and utterly risky true drama rivaled even the box office appeal of a *"Gone With the Wind"* ... that 1939 epic film with a story so huge and so personal that it captured the world's attention. Especially so since the penicillin storyline had an amazing climax of success.

The story did go on to capture the world's imagination. The unlikely lead character, incredible medical journal accounts, riveting clinical results, dramatic efforts to produce the drug and above all, the lives saved ... babies, young mothers, soldiers ... all added up to an amazing breakthrough phenomenon.

It seems that World War II at first interrupted British studies of the miracle drug and its potential for stamping out deadly bacterial infections. But then, enter the United States and American drug firms eager for other bacterial-killing drugs to aid or supplant the sulfonamides introduced in 1935. Such firms as Merck, Squibb and Pfizer feverishly worked to produce penicillin. By the time America entered the war, the nation's drug race was on: fighting men must have this drug. Widespread availability would mean life rather than death for vast numbers of suffering people.

The reader may look elsewhere for a detailed description of the search for ways to produce quantities of the precious, life-saving mold. By lucky accident, scientists discovered that America's corn fields yielded the perfect growth medium. England did not grow fields of corn, so penicillin supplies there had been scant and precious.

But in America, huge fermentation vats were built to supply the medium in which the mold cultures could be produced. Previously the process was described as a difficult fermentation, easily contaminated, with the penicillin yield disappointingly low. No one knew how to analyze the early results. It was therefore a slow scale-up process.

The wealth of the cornfields, in the American Midwest, the discovery of an even better strain of *Penicillium* mold and ever improving production methods, soon increased the penicillin yield many times over. The country's Office of Scientific Research and Development, in tandem with the American pharmaceutical companies, large and small, were producing increasing quantities of a drug so effective and yet so precious, that few citizens had access to it.

Dramatic recoveries from serious illnesses soon became known, and drug companies faced a difficult choice. Their human concerns and wartime patriotism often seemed balanced precariously by the huge corporate risks which would accompany attempts at large scale production. In the end, human compassion won the day as the Pfizer corporation's vice president and its Board voted to gamble everything on the temperamental, low-yield mold that was saving lives. The rest is history.

During those wartime years it became evident that penicillin could heal diseases that had plagued mankind from time immemorial. Demands for the wonder drug spread like a prairie fire on a windy day. By summer of 1943 the War Production Board recruited another twenty-one drug houses in addition to Merck, Squibb, Pfizer, Abbott and Winthrop, and in March 1945 commercial sales began. By then science had proved penicillin effective against gonorrhea, meningitis, syphilitic infections of the central nervous system and a growing list of others. These were not just palliatives, but definitive cures for previously intractable diseases.

The statistics were dramatic. Mortality from influenza and pneumonia fell by forty-seven percent in the United States between 1945 and 1955, deaths from diphtheria fell by ninety-two percent, and deaths from syphilis plunged seventy-eight percent. Infectious diseases, of course, had caused the great majority of deaths throughout history … until penicillin reversed the trend. What an extraordinary result from such ordinary beginnings!

Now picture, if you will, the dark-haired young man listening to *the* Dr. Alexander Fleming, discoverer of this penicillin, as he recounts quietly but confidently, many of these emerging facts to an enthralled audience in Turin, Italy in 1948. The discovery is still so new … results so thrilling … the work so inspiring … the future implications so mind-boggling … that everyone is in awe, including the speaker. A magnificent secret of nature revealed without fanfare to a quiet, unpretentious researcher.

That was a medical breakthrough, if ever there was one. It was one of a kind.

One medical student, Gustavo Bounous, listens transfixed as he turns over and over in his mind the enormity of what he is hearing. During those moments, his ambitions toward medicine and a newfound yearning for deep knowledge of the mysteries of the human body and its wondrous capacity for healing, spring into a tantalizing new life.

Who could imagine that an ordinary man like Fleming, an otherwise unremarkable worker in a mundane field of science, a researcher who plods away with little hope of any significant breakthrough … yes, even such a man could change the face of medicine.

Gustavo was more than awe-struck. He was transformed. Dante was right.

Here was a real man Gustavo could stretch his arm toward and nearly touch … a modest man with neither limousine nor chauffeur … one who didn't bother to pose with white coat and stethoscope … so average in appearance and deceptively unassuming …

Then Dr. Fleming dropped his final bombshell, the one that riveted Gustavo's attention.

"I didn't invent anything at all," Fleming concluded. *"My discovery happened by sheer accident."*

Did it really?

Gustavo returned to class and clinics a changed man. He was confirmed in his vocation. He now had a truly higher calling.

Years later, Dr. Bounous would reflect on Fleming's words with reverent gusto:

"I knew then, I was hearing a great man," he said. "He only *looked* ordinary."

4.

HOPE, GRIEF, AND HOPE RESTORED

The young man graduating in 1952 from the University of Turin as a doctor of medicine, certainly knew his own mind. At age twenty-four Gustavo Bounous had achieved the first giant step toward his destiny – ***he had become a doctor*** ! - and now he faced the next, even more difficult phase. He must move to Parma and then to Genova, where he would serve his four-year residency training in surgery. Yes, he would not disappoint his mother.

Thus Ada Bounous joined that fierce, proud company of mothers from every generation who see their beloved sons accomplish great goals against all odds.

Where medical school had required intense discipline, young Bounous was to discover that for him the process of becoming a surgeon in some ways seemed tantamount to working in a sweatshop. Those days he had precious little money for basic necessities, and certainly none for life's amenities. At an age where most young men relish getting out in the world, Gustavo Bounous had to grasp almost desperately for life's most basic needs: no fancy dining, no opera or *teatro* for him, no sports cars or lavish dates.

"The nuns used to feed me spaghetti," he admits. "They

were trying to convert me," he adds, mischievously. Dr. Bounous remembers the nuns and their hearty meals with gratitude. They catered to him with grace and the same adoration he was accustomed to from his own mother.

During his years of surgical residency, he found himself continually challenged to stretch his meager physical and financial resources.

"My training there cost zero, because I agreed to be on call every night in exchange for my bed and my spaghetti suppers." Sometimes he'd be awakened two, three, even six times in one night. Hastily throwing on his greens and scrubbing in haste, he would find that rushing to the emergency room, surgical suite or the operating theatre to assist, soon became routine.

"Those days I longed for sleep. I was always exhausted. Otherwise I believe my studies would have been even more successful."

Despite the non-stop pace of those long months and the deadening cumulative fatigue, Bounous managed to stay within the top ten percent of his class. He was gifted for sure, but his success did not come easily. Each day became the kind of grueling, march-to-a-goal routine that summons forth every fiber of one's determination and perseverance. Gustavo found the stuffy little basement room with its old bed, sticky heat and horrible humidity especially hard to endure. To be awakened repeatedly from a sweltering sleep, arouse oneself and hurriedly report to assist the surgeon on duty … to return exhausted for a few more precious snatches of sleep … then to set out, bleary-eyed, on a full day of rounds, labs and course work … that was grueling torture.

But Gustavo knew everything depended on his ability to stay the course. Luciana and their aging parents needed his support. Perhaps that's why each night prior to exams Bounous invariably found himself too anxiety-ridden to sleep. Tossing fitfully in the hard little bed, he tried to argue himself out of the tension that always rode so heavily on his shoulders.

"She relied on me," he explained, referring to his mother. "I had always, so strong a sense of responsibility."

Ironically, the heavy responsibilities that drove Gustavo to persevere also led him straight toward a terrifying professional impasse. The closer he came to completing residency requirements and earning the coveted diploma in surgery his mother always dreamed he'd receive, the more clearly he saw what he previously only suspected - a 'nobody' had absolutely no chance of practicing academic medicine in Italy.

"You had to be a relative or family friend of some prominent doctor," he said, "The medical chiefs who gave those assignments always became wealthy, very influential men. They were the gate-keepers to the clinical profession. You needed extraordinary influence with them to obtain any job.

"Despite those facts, however, by then I knew that my life had a purpose," Dr. Bounous adds. "I was a 'nobody' who didn't even mind standing up to my own Chief on occasion and telling him what I thought."

He recalls the time when he believed a fellow resident had been treated unfairly. Gustavo refused to be intimidated and confronted their boss.

"It's the system," their Chief alibied, dismissing the complaint with a shrug.

"Then the system stinks!", Bounous heatedly replied.

Each day the young doctor's personal concerns mounted higher. The closer he came to acquiring his idolized diploma and becoming a surgeon, the tighter he felt squeezed in the jaws of "the system." Always overworked, perpetually weary and anxious, he now sums up his residency stint in one emphatic word.

"*Unbearable*!" Bounous growls, his voice filled with disgust.

During that crucial final year of residency, life seemed at

its absolute worst. But looking forward, the end of the journey was in sight. The goal was now within reach but then fate delivered to Gustavo Bounous news so shocking as to render his own unhappy circumstances almost insignificant. His beloved mother had been diagnosed with cancer.

"Ovarian cancer," her son explains. "Even today, the prognosis is usually poor. That cancer is usually diagnosed late. It is aggressive and almost always deadly, with little hope for a cure ...so we were devastated."

As usual, the family rallied around the woman who had always rallied them.

"My father spent all the money we had on some top consultants for her," Dr. Bounous continued.

Ada's prestigious well-dressed physician, despite his professional reputation, his limousine and his access to the newest procedures and treatments that medicine could offer at the time, nevertheless found himself powerless against the relentless disease. Gustavo observed his mother's terrible suffering, the lessening of her physical strength, the loss of her dignity and beauty to the inexorable ravages of a disease which, over fourteen endless months, robbed them of Ada's irreplaceable spirit and life.

"My mother suffered terribly," Dr. Bounous reflected. "I could do nothing for her. Nobody could do anything for her. After all my hard work and training, I was seeing that sometimes medicine really does not work ... it is futile.

"My poor, gallant mother died in agony. I had to ask myself, what had all my efforts toward the study of medicine really accomplished? Were our sacrifices worth anything at all?"

Ada had seen her dream coming true. Her Tito was indeed a doctor – almost a surgeon – but she would not live to enjoy the benefits. She died prematurely.

Gustavo, the real mother's boy, had lost the prize. Or, was there another?

Work, even thankless work done under unpleasant conditions, provides an antidote for such tormenting questions. So Dr. Bounous redoubled his dedicated efforts. Gradually his deep grief over his mother's untimely death necessarily became superceded by urgent questions regarding his immediate future. As the final days of his residency program approached, he could no longer avoid facing the cold, unyielding stone wall of the unknown.

Where could he go? How would he find work? In fact, would it be possible for him to practice any medicine in Italy? He was in his late-twenties, responsible for supporting his family, and now a doctor, almost a qualified surgeon ... and he had no idea.

Gustavo's most nagging problem, ridiculously enough, was that he would soon have to give up that infernally hot, hateful room with its lumpy, uncomfortable bed. An incoming surgeon-to-be would need it. Therefore, time was of the essence. He must find a plan, and that he would have to do sooner rather than later.

His anxieties rose sky high as he considered his dilemma. Well-trained, well-liked and well-recommended as he might be, Dr. Bounous nevertheless saw himself as a doctor who might not have anywhere to practice, a surgeon who might find no place to operate. The yearned-for career which for years had shimmered before him and led him on, now appeared ready to die before it could take its first breath. He was facing his own still-birth experience.

"Bounous", his Chief hailed him one day, as the suspense grew nearly unbearable. "Aren't you the same one who always talks of wanting to practice medicine in the U.S.? How would you like to work in America?

The question hit Gustavo like a thunderbolt.

"Of course," he answered incredulously. "But how?"

"It can be arranged." the Chief countered. "I happened to tell a fellow in a place called Indiana about you, and he is willing to offer you a hospital position."

Dr. Bounous' jaw fell wide open. It seemed too good to be true. Was his mother an active angel still at work? Was there some sinister plan? Or, was proven excellence to be rewarded with recognition?

But how could Gustavo find money to make the trip? The cost must be enormously expensive. And what about the need to speak English? His French and Italian would not go far. "*It's all over, boys*," that American expression he'd heard the G.I.s shout from their Jeep that day so near the war's end … those few words wouldn't take him very far if he were actually to move to America … and those were almost the limit to his yankee vocabulary.

Gustavo stood before his chief, his mind leapfrogging and woolgathering at the same time, as he tried to absorb this startling new possibility. Hope leaped to life in his heart as his boss spelled out further possible details. There was already a plan of action if he would cooperate. A Genova doctor suffering from bladder cancer needed to send a tissue sample to a New York clinic …

Yes! Bounous could deliver the sample. Hand delivery was much the safest way ...

The same ill doctor would provide the promising Dr. Bounous with a one-way airline ticket to New York in exchange for his trouble. From there, he could make his way further West.

Yes! He must make arrangements to leave with all due haste ...

America! America! Dazed and somewhat disbelieving, his mind still afraid that there might be some undiscovered flaw in the plan, Gustavo at last allowed hope and joy to flood over his soul. America! And Indiana, at that … close to the land of his hero, Abraham Lincoln.

This incredible door of opportunity in his life, Gustavo decided, could only have come from the hand of God. Or rather, he amended, his *mother's* hand, no doubt guiding that of his Creator, or vice versa.

Just then his own hand was being pumped vigorously by his excited, very pleased Chief ... a wonderful fellow, Gustavo had always liked him ... as the two of them laughed and exclaimed together about the felicitous chain of events.

"And another good thing is," his boss intoned, "that some other fellow can have your excellent room. We really do need that bed."

On a cold day in February, 1957, Dr. Gustavo Bounous departed for Nice, France, from whence he would go west, cross the broad Atlantic Ocean, and arrive in America ... exactly as the boy Tito had dreamed, perched high in the branches of his chestnut tree. Was this a dream or what?

Something of that dreamy boy must have surfaced as the young surgeon embarked on the first and most exciting plane trip of his life. In the year to come, jet transportation would become common, but no matter, Gustavo liked everything about this plane, just as it was ... especially the blonde, long-legged stewardess who seemed so concerned for his comfort.

After a fuel stop in the Azores, the giant aircraft at last swept over the vast, unfathomable Atlantic, bearing at least one of its passengers to a new home he felt fully prepared to embrace.

Then, America ... Arrivals ... and U.S. Immigration, where scant attention or patience was available for a near-penniless foreign doctor from a similarly poverty-ridden Europe.

"It took four hours, with much paperwork and very little respect," Dr. Bounous recollected.

Then, at last, New York City. That incredible skyline actually did exist, looming incomprehensibly large and thrilling. Dr. Bounous delivered the tissue sample to the clinic as planned. Relieved at completing his unremarkable mission, he was eager to proceed to the New York Grand Central Station. First, though, he'd find something cool to drink.

Turning in to the nearest sidewalk restaurant, he uttered one of his few American phrases: "Coca-Cola."

"Huh?", the waitress responded, following with questions he could not understand. "Coca-Cola," Gustavo repeated more distinctly this time.

"Then came more questions," he related. "Now I know she was asking me 'what size?' But back then, we couldn't communicate at all. I gave up and had a drink of water."

It cost some thirty dollars to buy a train ticket to Indianapolis, Indiana ... a ticket to the heart of America, close enough to Lincoln's Illinois, and of course, in the mind of Dr. Gustavo Bounous, closer to something even more immediate. He was stunned as he saw his young life unfolding before his awe-struck gaze.

It must have felt good to spend one's last dollar on a ticket to one's future ... a young man's dream, and his destiny.

5.

AMERICA !

Another young man in Gustavo Bounous' situation might have felt intimidated. He was now far from home, alone. He was unable to speak the flat, uninflected language of midwest America. He was faced with learning a new demanding job in the strange city of Indianapolis, Indiana. He was among kind but unaccustomed colleagues. He had limited means at his disposal.

Gustavo Bounous, M.D., however, was no ordinary expatriate. Though he had imagined this very place, mid-America, for half his life, he had not permitted himself any particular expectations. Now he brought his energy and enthusiasm to a blank canvas, a fresh background against which he would begin to paint the panorama of his American future.

Various challenges loomed at once. It took time, for example, simply to find a room near the medical center that he could afford. What a pleasure eventually to discover it though, with its plain furniture, clean walls and yes, a sturdy bed. The contrasts between ancient, picturesque Parma and Genova, Italy and modern fast-moving Indianapolis were indescribable. Everything about his new life changes exhilarated him. There was more excitement than charm in this new modern metropolis.

To his surprise, at the Indiana University Medical Center, Bounous discovered that a state law prohibited foreign doctors from practicing medicine – and in his case, surgery. This, in the land of free-enterprise. He could hardly believe it. Yet, as if that were not dismaying enough, it seemed that his surgical chief in Italy had intended that Bounous learn the demanding intricacies of the heart-lung machine which was just coming into use.

"I was supposed to learn how to use this marvelous machine, then return to Italy to teach my chief's pupils to use it," he said. "Of course, in Italy I would never be permitted to do actual heart surgery – only the bosses performed surgery – but I would be allowed to teach the new techniques to them," he added, with some irony.

Dr. Bounous discovered that in America all doctors expected to receive hands-on training, with the surgical chief training and supervising those younger, rising surgeons.

"In Italy I held retractors or handed scalpels to the operating surgeon," he said. "My role as a surgeon could then be compared to that of a nurse assistant in America."

Dr. H.B. Shumacher, Jr., the Indiana medical center director, assigned Bounous to a research post. His project would tax his limited skills in dissecting laboratory animals. Experimenting with dogs, he began studying the circulatory system.

"Since I had had no actual practical training, even though I had already had four years of formal surgical training in Italy, I had to teach myself," he says.

He remembers the day when he first slowly, but carefully, tried to dissect a dog on his own.

"That was a nervous ordeal that I will never forget. I was anxious but excited.

"I was trying to slit a blood vessel, - it must have been an artery - and as you'd predict, I slipped and cut the wrong one," he recounted. "Blood immediately spurted everywhere…all over me, all over the laboratory, even onto the ceiling … it was a real mess."

Just at that very moment, a doctor whom Dr. Bounous had not yet met entered the room and surveyed the scene. The surprised Dr. Harold King was one of Bounous's new senior colleagues. It seemed the worst possible moment for the two men to meet. The older man, shaking his head in apparent disapproval, advanced toward the carnage. His younger colleague's heart began to sink.

"Dr. Bounous, I presume?", King asked with wry humor. The two men laughed, the tension broke, and King proceeded to show the young doctor how to perform the simple procedure.

"This surprised and relieved me very much," Bounous says. "He kept his composure. He accommodated my inexperience. I also found that everyone else was willing to take time to teach me anything I needed to know."

Such efforts on his behalf bespoke professional respect and acceptance. That produced a democratic atmosphere very different from his earlier experiences among the Italian medical hierarchy.

"In Italy the boss determines everything. It happened that my two supervisors there were Sicilians," he explained, "and Sicilians are very sensitive. There, the scene I just described might have been very embarrassing and unpleasant for me."

North Americans, Bounous quickly saw, were usually quite willing to blend professionalism with camaraderie. He thrived under his colleagues' helpfulness and enjoyed his new avenues for research and training. For the first time, too, Bounous was earning money. Even the small stipend he received before Dr. Shumacher secured a research grant for him, soon made Bounous feel a bit more secure and hopeful.

"In America I learned to drive a car, and then I bought my first automobile – a Ford," he recalled.

Everything, even the new language, seemed to come to Bounous far more easily those days. He learned English quite effectively from television. That was the decade in which TV began to invade more and more thousands if not millions of American living rooms. Bounous would hurry from work each evening, as did many American men, to switch on the novel entertainment medium.

"I especially liked the cartoons," he said. "Those cartoons taught me to speak English. My colleagues told me I spoke exactly like Bugs Bunny!"

But young, bright Bounous was no Bugs Bunny. He soon proved that.

Bounous's four-year stint at the Indiana University Medical Center found him working hard and accomplishing much. He acquired a grasp of English so quickly that within just eight months the novice Research Fellow was able to write creditable professional papers.

With the guidance of his new mentor Dr. Shumacher, Dr. Bounous not only acquired new surgical expertise, but applied his skills to tackle challenging questions related to the patho-physiology of renal blood flow and hypertension. He focused on his friendly dogs and became a master at dissection.

In some classic papers (1 - 4) they were able to demonstrate that "the autoregulation of renal blood flow is caused by the presence of the relatively non-expansible renal capsule." In simple terms, the tendency of the kidneys to maintain constant blood flow, as much as twenty percent of the cardiac output, and that despite wide changes in arterial perfusion pressure, is independent of external influences such as hormones and nerve activity. Rather, they showed (and others later confirmed) that removal of the kidney capsule was effective, in principle, in increasing kidney function in diseased states such as acute tubular necrosis or hemorrhagic shock.

Dr. Bounous also discovered that "systemic hypertension following experimental renal artery stenosis (renal hypertension) is aimed at maintaining a normal blood pressure and flow in the renal tissue: he coined the phrase *the egotistical kidney*. Again in simple terms, he found that if blood supply to either kidney was choked off, the kidney would trigger a rise in the animal's total blood pressure in an attempt to satisfy its own needs.(5, 6) For a modest man, this was the epitome of self-serving interest (organic avarice) that nevertheless, could be physiologically exploited by nature to serve the whole organism.

The surgical expertise that Bounous acquired while working on these dogs would later prove very useful, as we shall see.

In 1960, as he neared the end of his four-year term in Indiana, Gustavo Bounous suffered another major loss. He received word that his dad, Carlo Bounous, had died of hepatitis.

"My family did not tell me until after my father was buried," he said. "I could not return to Italy and then get back to the United States. The immigration rules would not allow me to do that."

In fact, his student visa would soon expire. He knew he would have to leave America then, after which it would be another two years before he could be permitted to apply again for new immigrant status.

Once more, it seemed, he must move on so another surgeon could occupy his bed, figuratively speaking, at least. Oh well, it would be worth the delay and inconvenience if he could return. But where would he go to in the interim?

Gustavo Bounous, M.D., by now wished with all his heart to become a citizen of the United States of America. He felt at home there. He was growing, thriving and producing in an enterprising atmosphere.

"Everything about my life has always happened by accident," Dr. Bounous reflects. "Because I had to leave the U.S., Dr. Shumacher took the trouble to secure a job for me in Canada. He had personal contacts at McGill University in Montreal, Quebec."

Reluctantly, the young researcher prepared to leave. The United States' growing involvement in Vietnam, with the concurrent U.S. – Cuba turmoil, now made it impossible for the doctor to extend his visa. In those days, many political refugees and thousands of others seeking asylum from revolution and repression elsewhere were streaming into the United States. Since Bounous's life was not in danger, he had but little claim to a coveted immigrant status. Once again, the world's wartime and political circumstances were to determine Gustavo Bounous's direction and goals.

"I always take the path of least resistance," he philosophizes. Again, it appeared, he had no other choice. "I had to leave America ...if I returned to Italy, I would have no job ... yet there in Montreal, at the prestigious McGill University, I could continue my research work in their Department of Experimental Surgery."

Disappointed at having to leave the U.S., yet grateful for his new opportunity, Dr. Bounous could never have guessed how fortuitous his new appointment would prove to be.

With mixed emotions, Gustavo Bounous obtained a visa and set his face toward Quebec.

"In Italy I was a nobody, but in Indiana I found the cradle of human kindness and respect," he reflected. "I had no idea how it would be for me in Canada. I did have initial misgivings."

Once again he must learn to fit in among new teachers, researchers and surgeons within one of the world's premier medical institutions. At McGill he would go forward with his research on hemodynamics, the intricate mechanics of the human circulatory system.

Dr. Shumacher, his chief of surgery at the Indiana University Medical Center, had distinguished himself within the field of heart surgery. Bounous knew that his new mentor was at the forefront of recent advances. Therefore with Shumacher as his sponsor, Bounous's position within McGill's medical research structure appeared quite promising indeed. He now had some academic leverage.

However, one could hardly say the same for the young surgeon's personal life. At thirty-four, unmarried, with both parents now dead and his sister Luciana to support, Gustavo Bounous also found himself literally a man without a country. His only ticket to the future, a temporary visa into Montreal, Quebec, must have seemed flimsy indeed to the man who so ardently preferred to remain in the United States, the land of Lincoln. Little could Bounous imagine, however, how critically important to his life, and to his future in medicine, Canada would become. In fact, Bounous's destiny (he later learned) was to reside not in the United States, but in Canada.

The young man could not possibly foresee, as he arrived in Montreal, poor and dispossessed, that within three years' time he would achieve a signal honor. Gustavo Bounous, M.D., would become the only non-Canadian to date ever to be awarded the distinguished Medal of the Royal College of Physicians and Surgeons of Canada.

1958

“Luck is infatuated with the efficient.”

- Persian Proverb

6.

BORDER CROSSING

The course of many a life is flooded with blessings in disguise. And so it was for this young devoted medical researcher.

If Gustavo Bounous summed up his surgical residency in Italy as providing him little at the time beyond hard work, privation and a professional dead-end street, he also acknowledged that it became the unexpected vehicle which catapulted him into prestigious opportunities in the United States and Canada.

Bounous had not only left the land of Italy behind. He had escaped the childhood of fascism, the stifling elitism of the establishment he had known and his own personal lifestyle of frustration. He readily embraced midwest American life and more, and soon decided to make this new world his permanent home. He gained a new appreciation for the values of hard work and entrepreneurship, of freedom and reward. He was seeing at least this version of Americana close-up. And he loved it.

"In Indiana I was beginning to learn what democracy is … about the nature of human relations and respect … and what was required to develop good skills in experimental surgery. I felt optimistic and happy there," he said.

Dr. Bounous revered his American boss, Dr. H.B. Shumacher, Jr., not only as a great surgeon but also a great gentleman, one who believed in the young aspiring surgeon's talent. For his own part, Bounous considered himself extremely lucky to have the chance to work so closely with one of America's pioneer heart surgeons. As the younger man's skills and confidence increased, he quickly began to make real contributions to the cardiovascular research developing through Shumacher's leadership.

Those four years, though in some ways perhaps even more challenging than his difficult years of training in Italy, produced in Bounous not only steady professional growth but some priceless personal rewards, as well.

For one thing, Bounous blossomed in America's democratic atmosphere. He liked his colleagues' openness and lack of professional posturing – no limousines here with ostentatious symbols on the license plates – and the genuine, open natures of the people he met. Their characters often brought to mind his mental image of America's Abraham Lincoln, as described by Carl Schultz in 1864:

" ... his manners are not in accord with European conceptions of the dignity of a chief magistrate. (Or the dignity of some Italian physicians, Bounous might have added.) *He* (Lincoln) *is a well-developed child of nature and is not skilled in polite phrases and poses. But he is a man of profound feeling, correct and firm principles and incorruptible honesty. His motives are unquestionable, and he possesses to a remarkable degree the characteristic, God-given trait of this people: sound common sense."*

McGILL

Dr. Shumacher's own great common sense had led him to obtain for his protégé, who by now had become a valuable and dependable member of his research team, a post at the McGill University Medical Centre. There, as an assistant professor in the

Department of Experimental Surgery, Bounous could continue hemodynamics experiments under the direction of the eminent Canadian surgeon, Dr. Fraser Newman Gurd. After his two years outside the United States, the men reasoned, Bounous could then apply for U.S. citizenship, return to Indiana, and resume his contributions to the university's impressive body of medical research data. Thus, crossing the U.S.-Canada border appeared to be an almost seamless career change for Bounous, one the trio of surgeons believed would prove mutually beneficial.

It's easy to imagine that Montreal, Quebec became something of a spiritual bridge between the classical, historic and picturesque Italy of Dr. Bounous's youth, and the energetic enthusiasm he had discovered in the much newer America. Montreal's cosmopolitan population, its proximity to mountains and lakes, its occasional uneven, cobblestoned streets, not to mention the multi-lingual conversations one overhears in shops and restaurants, felt more than a little like Europe. It was a city of culture and charm.

However, Montreal's sprawling suburbs, thriving manufacturing plants and above all, the thousands of other individuals like Bounous who arrived in search of jobs, asylum or a fresh start in life, reminded him of a teeming, enterprising America.

The serious-faced young man who approached the cavernous McGill center as a newcomer to Canadian medicine may have felt somewhat apprehensive. The edifice itself, a rather foreboding seventeen-story pile of dark red brick with a gothic-style, granite-faced entrance, looked large enough to house the population of a sizeable village.

Established in 1829 with a Faculty of Medicine as its cornerstone, McGill University had long enjoyed a worldwide reputation for scholarly achievement and scientific discovery. As Dr. Bounous climbed the steep, stone frontsteps, pushed through one of the enormous front doors and joined a bewildering mass of humanity – old folk in wheelchairs, white-coated resident physicians, nurses in starched white caps which resembled winged birds, young mothers with crying infants – he could not foresee that within these vast clinical surroundings he would achieve his first internationally recognized medical discovery.

That first day, as he hurried through patient-lined corridors that seemed to stretch almost to infinity, Bounous's mind fixed on the person of Dr. Gurd. What sort of individual would this Gurd prove to be, he wondered? How would he look upon a surgeon who had trained in Italy? And, he couldn't help wondering, would he be fortunate enough to receive the kind of friendliness and respect at McGill that he had so enjoyed in Indiana?

Dr. Bounous need not have worried. In Fraser Gurd, he discovered the same unassuming and caring personality Shumacher had displayed. He was given a warm and friendly welcome. That put Bounous almost immediately at ease. He knew he was with good company, not just a distinguished professional.

"Truly great people always possess humility," Bounous remarked. "They are interested in you and want to help you."

Dr. Gurd, white-haired and courtly, immediately took the younger Bounous under his wing, displaying an avuncular interest in his welfare. Bounous reciprocated with gratitude.

"I wanted to work hard for him, to repay his many kindnesses any way I could," he said.

RESEARCH LIFE

Sir Winston Churchill's famous phrase, "*a riddle wrapped in a mystery, inside an enigma,*" might well describe Gurd's and Bounous's challenging work with hemodynamics. At McGill University, as at the Indiana University Medical Center, Bounous became vitally involved in the dauntingly complicated study of blood circulation, and all that pumps, filters, clots, flows, feeds, pressures, obstructs or provides the heme (blood) which pulses through veins and arteries and threads its way even through the tiniest capillaries.

Bounous's surroundings at McGill provided a stark background for such drama. Situated midway along a ninth-floor corridor which fed into a series of small, severe offices, Dr. Bounous had been assigned his own special cell within a honeycomb containing some of the world's brainiest, most questioning medical scientists. For decades, countless footsteps of some of the medical great had imprinted that same brown-and-tan rubber-tiled corridor that led to Bounous's stuffy, crowded cell.

Not that he thought of such things, of course. Bounous, by nature, training and experience, and now a thoroughgoing scientist, was not seeking drama, but medical truth. Industrious, patient and observant, he fit easily into the disciplines of pure research. That called for a subtle dynamic of first question, then premise, followed by method, practice and observation. After early information gathering, would come analysis, then more questions, more practice and observations; then finally, careful recording of everything and hopefully, eventual proof to cap the exercise. That is the anatomy of research.

Of course the Bounous scientific mindset had been largely formed long before his medical training commenced. As a boy dreaming within the branches of his chestnut tree, he began devouring the writings and absorbing the thinking of those immortals of the Italian Renaissance who had bequeathed him their dazzling discoveries in such fields as art, letters, politics, astronomy, science and medicine.

Surgery itself actually began in Italy. And though the boy Tito never dreamed in that direction – recall that literature would have been his choice – now, entering his thirties, he was following in the footsteps of the famous sixteenth century pioneer surgeon, Faré. Interestingly enough, he noted that the earliest surgeons were not physicians at all, but skilled physical "mechanics", as it were, men dexterous enough to perform actual surgical procedures as the physician in attendance directed them.

Investigation, data-gathering and assessing evidence became Dr. Gustavo Bounous's life. During those early years, one Maxwell Maltz, M.D., an American plastic surgeon, wrote the best-selling book, "Psycho-Cybernetics", which described the human brain as a goal-setting mechanism, designed to seek out and move toward a continuing stream of goals or challenges to attain optimum function. Maltz's vivid description of the human brain's mechanics and the ways in which one should "set" one's mind for best success, largely describes the very methods by which Bounous and other medical investigators went about their work.

"*Conscious, rational thought selects the goal, gathers information, concludes, estimates, evaluates and starts the wheels in motion,*" Maltz wrote. "*It is not, however, responsible for results. We must learn to do our work, act upon the best assumptions available, and leave results to take care of themselves.*"

Exactly. Most research scientists, including Bounous, explain that data resulting from even the most painstaking investigation very often yields no tangible or apparently useful conclusions. As Maltz further wrote,

"*We can have no intimation or certified guarantee in advance that it (the investigation or goal) will come up with the answer. We are forced into a position of trust. And only by trusting and acting do we see signs and wonders.*"

Few lay people can imagine a life consisting of endless scientific inquiry; tedious hours of observation and record keeping; the laborious writing of proposals, reports and scientific papers – all with little or no expectation of ever unearthing a startling medical breakthrough.

But that is the life of the true researcher, the pure scientist in search of truth as revealed in nature. It is a process that justifies itself; one that more often than not, will lead to much of little consequence. Yet there is always the remote possibility that without warning or prediction, profound results can arise when least expected. The phenomenon is known as *serendipity*. It is never a goal, it is only by chance.

Such things do happen. Recall Sir Alexander Fleming, who discovered in this century, the mold that killed bacteria. That discovery, now years later, has become the commonplace 'miracle drug', penicillin. Landmark medical discoveries by two famous nineteenth century scientists, Pasteur and Lister, changed the course of medicine forever, as did the work of the men associated with general anesthesia: Long, Morton and Warren. And those are just a few examples from medical science history.

On the way to achieving miracles, or even lesser but still valuable medical discoveries, mountains of isolated fragments of valid research and experimentation continue to pile even higher. Sometimes the tiniest particles of truth may later connect with another man's work – perhaps decades later and half a world away – with astoundingly meaningful results.

So Gustavo Bounous became part of that endless parade of the relatively few men and women in each generation who dedicate themselves to the pursuit of medical knowledge. His published research papers treated such subjects as blood coagulation, arterial diseases, vascular disease in the elderly, experimental observations of the use of a mechanical bypass pump during cross-clamping of the thoracic aorta, and the value of angiography in the evaluation of vascular disease – all pioneering study at that time. Other investigations dealt with the use of nylon mesh for esophageal support; further studies of renal blood flow and autoregulation; cardiac output and blood pressure; and renal flow response to occlusion of visceral arteries.

These and other similar published works written in collaboration with Dr. Shumacher and two or three other University of Indiana colleagues in surgical research helped solidify

Dr. Bounous's growing reputation prior to his move to Montreal, and served to forecast the seminal experimentation he would conduct with Dr. Fraser Newman Gurd.

Gurd, it seemed, intended to lead some studies into hemorrhagic shock. The challenge: that of discovering why certain hemorrhaging patients, despite receiving all known appropriate fluid therapy, could not be saved. All Bounous's previous work had prepared him for this reasearch.

Some time before he crossed the border into Canada, however, Dr. Gustavo Bounous achieved a far more significant border crossing: that of changing his self-perception from that of a "nobody" to a far more realistic picture of the well-respected, dedicated scientist he had become. Now, together with McGill University's highly regarded Dr. Gurd, Bounous had one more figurative border to cross. This one would lead him and his distinguished research chief into sudden, high-profile recognition and honor.

7.

YEAR OF THE DOG

To induce hemorrhagic shock in a laboratory animal for purposes of scientific study must require heroic amounts of detachment, determination and exquisite medical research skills on the part of the surgeon. During his first two years in Canada, Gustavo Bounous effectively acquired this special mix of traits. Dr. Gurd, his Chief, recognized the young man's brilliance early-on and encouraged him to tackle ever more difficult problems.

Their challenge, hemorrhagic shock, was not only daunting and perplexing, but exceedingly messy to investigate. Dogs presented the laboratory specimens of choice, since canine hemodynamics so closely resemble those of *homo sapiens*. To cause in the animal the copious internal hemorrhage which would eventuate in extreme shock usually took some three hours, during which every reaction would be closely monitored and carefully recorded.

Dr. Bounous, working most of the time with only the aid of a female under-graduate technician, was to observe and time each all-too-predictable stage of the massive, systemic shock. There was copious bleeding, accompanied by an ominous drop in blood pressure which, despite quantities of replacement fluids, almost inevitably resulted in death. Why? That was the sixty-million-dollar question.

Every physician in the world confronted with massive hemorrhage and the victim's resultant profound, usually fatal shock was acquainted with the grim sequalae. Too often, in an emergency scenario, despite all due speed, skill, prompt fluid therapy and heroic medical efforts, the patient died. Doctors Gurd and Bounous asked: 'What controlled the sinister unresponsive drop in blood pressure, the oxygen deficiency which predictably gave the skin a bluish tinge, and inevitably, the steady and often rapid shutdown of internal organs?' That question was definitely worthy of investigation.

Dr. Gustavo Bounous, at Dr. Gurd's request, became the investigator of choice. Time after time he opened a profusely bleeding animal, now in profound shock, searching for some metabolic or physical sign which could provide any probable answers to basic questions surrounding the mechanisms of hemorrhagic shock. He met with repeated failure. He found nothing to reveal an essential clue.

SERENDIPITY

His great breakthrough, as often happens, occurred 'by accident.' As Dr. Denis Waitley, the American psychologist, has observed, "*Luck lies at the intersection of preparation and opportunity.*"

On one fateful morning, unannounced and unrehearsed, Dr. Bounous had arrived at that intersection. His 'luck', ironically, consisted in his being given a single animal which had not been fasted, as was the usual protocol. Because he was prepared by due diligence, that morning he intersected a great opportunity.

"I was interested in what happens to the internal organs, particularly the intestines, during hemorrhagic shock," Dr. Bounous explained. "Prior to such experiments, our animals had always been fasted. But surprise, surprise! That morning, someone made a mistake. One single animal had been fed. Why? Who knows. What I do know is that in this singular case, there was a very poor response: the shock occurred very quickly! I opened the abdomen … then opened the intestine … and discovered …

tremendous hemorrhaging. There was no ambiguity.

"I was surprised and mystified, and still unaware that the animal had been fed," he continued. "The intestine looked very bad ... different from other such specimens. When I examined a segment of tissue ... I saw that the intestinal wall had been worn thin!"

Since the hemorrhagic denuded tissue had been immediately distal to the pancreatic duct, Dr. Bounous instantly surmised that the digestive process somehow had affected it.

"I guessed that probably trypsin from the pancreas must have eaten through the intestinal mucosa and wall, allowing poisons from the gut to enter the body, thus creating irreversible shock."

He was right. In a subsequent experiment, local pretreatment with a pancreatic protease inhibitor prevented the development of hemorrhagic necrosis of ileal mucosa in a dog during severe shock.

The moment of discovery electrified Bounous and his young assistant.

"I took a piece of the affected gut without hemorrhage and ran down the hall to the old fellow who did our scientific photography," he recounted. "After he photographed the specimen I literally ran to Dr. Gurd's office. He became as excited as I was. We both knew instinctively that this would be a very significant new finding."

As we have already noted, scientific research is a long and arduous exercise of daily discipline and devotion to the search for nature's truth. It is painstakingly adding one piece of a giant jigsaw puzzle upon another, with no master blueprint available. The process is one of calculated trial and error that is otherwise called experimentation. Day in and day out, the researcher designs and prepares different controlled conditions. Then he or she observes and records results, both positive and negative. It is often lonely and unrewarding, but then ... at those rare moments of discovery, the routine is interrupted by ecstatic moments of delight.

Even the young but mellow Dr. Bounous could not contain his excitement that day.

The landmark 'accident' proved to be not only a scientific breakthrough, but a landmark event in Bounous's career, as well. In the 1964 *Annals of Surgery*, Bounous reported the finding with doctors L.G. Hampson and F. N. Gurd: "***Cellular nucleotides in hemorrhagic shock. Relationship of intestinal metabolic changes to hemorrhagic enteritis and the barrier function of intestinal mucosa.***" That was a watershed paper.(7)

THE MEDAL

Thus, at age thirty-six, Bounous's steady focus on hemodynamics had already led him into precaucious scientific achievement. The paper describing his discovery meanwhile made its way to the attention of the eleven-man committee charged with selecting that year's recipient of the prestigious Medal of the Royal College of Physicians and Surgeons of Canada. This meant not just respect for hard work or diligent study, it was the consummate recognition from his peers.

In January, 1965, Dr. Gustavo Bounous was summoned to Toronto to be honored at ceremonies wherein he received the coveted medal. Luciana, his sister, flew from Italy for the momentous occasion. She would personify their mother's pride in her brother Tito.

Not only was Luciana's brother singled out for the important award at an unusually early stage in his career, but he also had become the first non-Canadian physician ever to receive it.

Anyone takes pleasure in becoming Man of the Year, as it were, in one's profession. Typical of Dr. Bounous, however, was the fact that he took an even more solid pleasure in seeing his professional accomplishment also enhance his benefactor, Dr. Gurd. The two men had worked closely together, and Gurd deservedly shared Bounous's spotlight. Soon Dr. Fraser Newman Gurd was promoted to head of the Department of Surgery and no one could have been more pleased

The Royal College of Physicians and Surgeons of Canada

Office of the Secretary

PATRON: HER MAJESTY THE QUEEN

74 Stanley Avenue,
Ottawa 2, Ontario

November 18th, 1964

Dr. Gustavo Bounous,
University Surgical Clinic,
The Montreal General Hospital,
Montreal, Quebec.

Dear Doctor Bounous,

It is with pleasure that I inform you that the Committee on Annual Awards has selected your essay entitled: "Cellular Nucleotides in Hemorrhagic Shock" for the medal of the Royal College of Surgeons of Canada for 1965.

The medal..., will be presented to you at the Annual Convocation ceremony which will be held at Convocation Hall, University of Toronto, on Thursday evening, January 21st, 1965, at 8:30 p.m. You will be in the academic platform party on this occasion for which the dress will be academic gown over dinner jacket (black tie). The College will furnish a gown for you.

It is probable that representatives of the press will wish to interview you on the subject of your essay. This would be arranged through Dr. Neil Watters who is Chairman of the Press Relations Committee for Surgery. In any case you should furnish this office with two copies of the text of your paper as it will be presented....

Further details and instructions will be given to you by the undersigned when you register at the meeting.

Yours sincerely,

T.J. Giles
Executive Secretary

for him than his own protégé, Gustavo Bounous.

"After I received the 1965 medal, everyone was so willing to reward Dr. Gurd," Bounous recalled. "I was delighted when he was promoted to Surgeon-in-Chief. He was such a nice man, and so proud of me. He even presented my work in the United States."

Like Cinderella returning from the ball, Bounous returned to his small, plain research cubicle. Although he had been hired as an Associate Professor of Surgery at McGill University, he says he really never much enjoyed stints in the classroom. He much preferred the hands-on work of surgical research to that of lecturing, and while he always taught a few classes throughout most of his career, he made sure he kept the demands of academia to a minimum.

Not long after, Bounous published other papers following his now-famous discovery pertaining to hemorrhagic shock – "*Abolition of tryptic enteritis in the shocked dog. Creation of a model for study of human shock and its sequalae*(8)"; "*The cessation of intestinal mucous production as a pathogenic factor in irreversible shock* (9)", and "*Metabolic changes in the intestinal mucosa during hemorrhagic shock* (10)", all in the following year. Other researchers also reported the same intestinal effects resulting from radiation therapy.

Impaired capacity to maintain the protective mucous coat of the enterocyte during low blood flow states was found to be a factor of reduced intestinal resistance to pancreatic proteases and other noxious substances. For example, the intestinal mucosa, normally impermeable to curare, would then allow the passage of this substance.

The spectacular appearance of hemorrhagic necrosis in untreated mucosa contrasting with the area where pancreatic proteases had been inhibited focused the attention to the local phenomenon.

Indeed these remarkable intestinal lesions were "markers" of a wide systemic syndrome caused by the loss of intestinal barrier to luminal contents as well as release of toxic compounds from the damaged mucosa. For example, autoinfection with common enteric microorganisms is now recognized to be a leading cause of late morbidity and mortality after trauma, critical surgical illnesses and shock.

ELEMENTAL DIETS

The discovery of the importance of pancreatic secretions in the destruction of the wall of the small intestine when the protective mucosal barrier is removed, was much more than academic. There were obvious and important implications. If feeding the dog stimulated the digestive process leading to "digestion" of the wall, could the same wall be protected if the animal was fed 'predigested' foods which supported the mucosa and at the same time reduced the digestive enzymes in the gut? Dr. Bounous proposed such a prophylactic use of what he first called, an "**elemental diet**" (for want of a simpler or better term). This pushed him right into the mainstream of the nutrition field but at right angles to the medical establishment. The idea that one could use diet or nutrition to prevent and even manage such serious clinical conditions as hemorrhagic shock, radiation sickness, cancer chemotherapy or inflammatory bowel disease was a revolutionary one. But Bounous had become a pioneer before he realized it.

"Dr. Gurd and I brainstormed one day in his office," Bounous says. "We couldn't decide what to call this 'food-medicine'. At last, I decided to call it the 'elemental diet', and it is known by that name today."

An "elemental diet" (ED) is defined as a water-soluble formula diet containing essential nutrients mostly in their simple molecular form. Amino acids and small peptides (hydrolyzate) replace proteins; free fatty acids replace fats; and lactose-free mono-and di-saccharides replace complex carbohydrate. Minerals and vitamins are included to meet or exceed requirements. No fiber as such is present. These are all contained in a liquid solution for use by mouth or through an inserted tube.

The prototype diet used in 1967 contained fibrin hydrolyzate, sucrose, and small amounts of triglycerides.(11,12) It was observed that when dogs drank this diet exclusively for three days, and were then fasted overnight before controlled hemorrhagic hypotension or intestinal ischemia, the intestinal lesion was minimized in comparison to dogs pre-fed standard laboratory food. Moreover, in accordance with the role of the intestinal lesion in the pathogenesis of shock, survival was significantly improved.

Because the same types of injury can lead, in survivors, to a severe defect in the terminal digestion of nutrients, elemental diets, with minimal digestive requirements, are also effectively utilized as a therapeutic tool following multiple trauma, major surgery or severe burns, when insufficient terminal digestion in the brush border and in the enterocyte can be a limiting factor to absorption of nutrients.

These experimental data prompted Dr. J.E. Knapp of Mead Johnson Canada to develop a casein hydrolysate-based ED for human consumption. This diet, labelled 3200 AS, was found to be as effective as the original fibrin hydrolysate formula in protecting dogs from hemorrhagic shock when fed exclusively this ED during 4 days prior to injury. Subsequent studies in human volunteers showed that product 3200 AS, a nutritionally complete elemental formula containing casein hydrolyzate, was safe for human use. In 1972, product 3200 AS became commercially available as “Flexical”. This became the standard for the further clinical studies that Dr. Bounous undertook. There are now several commercially available ED’s including Criticare H.N., Vital, Accupep, Vivonex (USA), Tripeptid (France), Elental (Japan), etc.

Subsequent positive results were obtained in experiments with other laboratory animals and more importantly, in clinical patients suffering from hemorrhagic shock, severe burns, trauma, post-operative complications in intensive care, multiple organ failure, radiation enteropathy, chemotherapy, Crohn’s disease and other serious medical conditions. These dramatic effects of the “elemental diets” in both prevention and treatment exceed the noted benefits of intravenous feeding with a nutritionally equivalent formula or an equivalent normal diet containing whole proteins.

When the intravenous was compared to the enteral route, following major abdominal trauma it was clearly demonstrated that the complication rate and the overall cost of therapy were lower with the latter than with the former approach.(13, 14) Enteral feeding was thus proven superior to the intravenous route in most cases and elemental diets were better in cases with fragile intestinal epithelium and damaged brush-border.

No one was more surprised by all this fallout than Dr. Bounous himself. Who could have guessed that such a simple approach of modifying the diet of patients in this way could be so effective in the management and prognosis of these patients? The increasing widespread application of this approach in ICU's is clear testimony to its effectiveness (and at demonstrably reduced cost and morbidity). The therapeutic use of elemental diets opened the door to the wider concept of "enteral" nutrition as opposed to "parenteral" nutrition.

But 'what works' in true science does not always win out in the real world outside the laboratory. Many other practical and commercial factors dictate the fate of all the potential applications of any new development. Suffice it to note that intravenous feeding with costly pharmaceutical preparations is big business, with powerful economic interests and marketing forces. Enough said for now.

Step by step, the surgeon had moved from studies of cardiology to those of hemodynamics, metabolism, and now, nutrition. But it was more than that. As a proven outstanding researcher Dr. Bounous had made a major contribution in each field as he "passed through". Yet he was still questioning himself. Had he not ventured too far afield?

"Because of the 'elemental diet' I became recognized as a pioneer in nutrition," Bounous says. "Nutrition was not interesting to most doctors in those days ... it was not at all fashionable. But it seems I always followed my own path, seldom choosing the popular or glamorous research avenues to explore. I was only following where my research led me. There were new ideas that I could not contain."

His 'unfashionable' elemental diet was to become the forerunner of his yet future medical discovery which could hold far-reaching health implications. But that was more than a decade away. He had no premonition, only scientific curiosity.

Just ahead, though , Gustavo Bounous would travel a professional road so full of twists, turns, 'accidental' discoveries, frustrations, disappointments and frequent discouragement, that a less tenacious man might easily have sought a more popular or glamorous line of research.

Fortunately, Dr. Bounous never abandoned the rocky, often tedious professional road he had chosen – one he would traverse for another twenty plus years.

Midway on the journey he would once again stumble on a "lucky ...", "accidental ...", "amazingly fortuitous ..." leap of scientific discovery. It would be so sweeping in its potential implications for many a person, that no physician or researcher to date has probably fully comprehended its possibilities. It took Dr. Bounous himself years to even begin to appreciate what he found.

Quite simply, Bounous's unselfish style, his surgical research interest, both coupled with his growing knowledge and reputation of how nutrition affects the body at a cellular level, was eventually to attract the right components that would lead him into findings now emerging as some of the most significant medical discoveries of the Twentieth Century.

It's a truly fascinating story - one that began, at least in part, during that busy year, 1964: for Dr. Gustavo Bounous ... the Year of the Dog.

1968

“ … rejected of men”

8.

DISAPPOINTMENT AND DESTINY

From the high peak of professional success, Gustavo Bounous soon was to slide into a deep valley of profound personal dis appointment. The time to break his ties with Canada had arrived; the doctor must wind up his engrossing work in Quebec and prepare to emigrate back to America. Back to the land of Lincoln he would go.

Montreal, he reflected, had been very good to him. By now, Bounous very much appreciated the importance and high standing McGill University held in the world's medical community, and in scientific discovery in particular. And McGill had respect for his work and reputation. Indeed, he counted himself fortunate to have been part of the institution's robust research efforts. It was a good marriage but he had not forgotten his first love. He had never forgotten his American dream. Despite Montreal's livability, richness and charm, he still wanted to go back to the heart of the United States. He had stayed away long enough. It was time to return to America's freedoms and possibilities and become one of her adopted sons.

The day he presented himself at the U.S. Embassy in Montreal, however, the doctor's lifelong dream became abruptly shattered.

"You are not qualified to immigrate," the official told him.

"Of course! I left the country for three years, as I was told, and now I request to return and apply for citizenship,"

Bounous explained.

The verdict came, swift, blunt and overwhelming like acute physical shock.

"No, you are not eligible. You should have gone back to Italy, then entered the U.S. from there. You cannot emigrate from Canada."

Bounous attempted to argue, but to no avail. Stunned and defeated, he left the embassy with the official's words ringing in his ears.

"Sorry, sir, but you missed the bus."

News of Bounous's important scientific achievements, however, had crossed the Canada-U.S. border. On March 2, 1965, U.S. Representative Frank Annunzio of Illinois, entered the following remarks into the **US Congressional Record:**

> "Mr. Speaker,
>
> One of the most important problems we will deal with in this Congress is the urgent need to overhaul our entire immigration policy. Hardly a day goes by without my encountering some instance of the irrational and unjust manner in which our present immigration laws operate.
>
> "The March issue of *Fra Noi*, a monthly newspaper reaching a quarter of a million Italian-Americans in the city of Chicago, contains two articles on this matter which perfectly illustrate my point.
>
> "The first is a news dispatch from Toronto reporting a top medical award to Dr. Gustavo Bounous, an Italian immigrant to Canada. Dr. Bounous lived for a short time in America but was forced to move to Canada when his temporary visa ran out. Mr. Maurice R.

Marchello, an attorney in Chicago, uses the case of Dr. Bounous to show how the 'national origins quota system' works to the detriment of the United States.

"I recommend these informative articles to all of my colleagues; therefore, with unanimous consent, I insert them in the Appendix of the Record.

"The articles follow:

I.

DR. BOUNOUS STORY – WE LOST HIM

Toronto-*Montrealer Gustavo Bounous, dynamic example of brain at work in Canada's favor, received a top national medical award here last night for shedding a bright new light on how shock from bleeding causes death.*

Dr. Bounous' work, which won him the medal in surgery from the Royal College of Physicians and Surgeons of Canada, opens up the possibility of saving patients who succumb to hemorrhagic shock despite transfusion.

An Italian surgeon trained in Turin, Dr. Bounous, 36, moved to Montreal early in 1962 from Indianapolis when he was forced to leave the United States because his temporary visa ran out.

Working with laboratory dogs, Dr. Bounous tackled the puzzle of why early transfusion and treatments save patients in shock from bleeding but fail after a certain time that varies from person to person.

Dr. Bounous found that the chemical structures of cells in the lining of the bowel change in a state of shock until they reach a point where they can no longer accept oxygen. This cell exhaustion weakens the bowel lining, permitting poison body waste to escape and kill.

Putting dogs into a state of shock by loss of blood, he measured the ability of intestine cells to accept oxygen at intervals. He found that chemical changes gradually reduce the ability to take oxygen from the blood. The enzyme substance that permits the transfer of oxygen from blood to cell finally is depleted, causing complete breakdown of the cell.

If medical science can isolate and identify precisely what chemical is depleted in shock, it might someday be possible to give patients, who fail to respond to transfusions, a life-saving injection of this chemical.

II.

CONGRESSMEN MUST ACT NOW – BIASED IMMIGRATION POLICY IS BRAIN DRAIN TO UNITED STATES

(By Maurice R, Marchello)

We as Americans of Italian origin do not envy Canada's gain in acquiring a fine citizen, but we are deeply concerned with the loss to America of such exemplary citizen prospects, due entirely to the shortsighted laws which limit and restrict them from entering our country.

We respectfully submit that all Americans should be equally concerned with this "brain-drain" – especially our Congressmen, who now have the power to correct it.

One can only conjecture how great this brain loss must have been over the past years. Just try to imagine what a tragic loss of talent our country would have suffered in the past 30 vital years if the Italian-born Fermi, Fubini, and Conto – to mention a few outstanding scientists – had to cope with the quota system to gain admittance to our country.

Fortunately, these three brilliant contributors to our country, because of their special circumstances, were able to avoid the immigration quota law restrictions.

Enrico Fermi, the father of the atomic and nuclear age, was accorded political sanctuary in our shores because he was an anti-Fascist exile.

Eugene C. Fubini, the valuable Assistant to our Secretary of Defense MacNamara, and who controls our military communications satellite program was admitted outside the quota because his wife was an American citizen.

The late Chicagoan, Armando F. Conto, a television pioneer and a well-known figure in communications engineering, also was admitted as a non-quota immigrant because he was fortunate enough to have married an American citizen, whom he met in Europe when she was a student of foreign languages.

Congress now has the opportunity with presidential sanction to abolish forever the hypocrisy of our antiquated immigration policy. The quota system based on where a man is born has too often damned the gifted equally with the deprived.

American immigration policy should serve

> *the best interest of Americans. While specifically we are now pointing up the hard plight of the humble, talented Italian prospective immigrants, we ask no more for them than we do for others, regardless from whence they come.*
>
> *There is also the great need to correct, with compassion, certain wrongs done in the past. The agonizing separation of families brought about by the old quota system also is a strong consideration for the passage of the proposed legislation.*
>
> *Basically, the recommended legislation gives equal treatment to all. So let us now search not only in the heretofore favored lands of the Nordic Anglo-Saxons, but let us also climb the hills of Rome and Turin and explore the Mediterranean shores where, in the future, perhaps another Fermi, Fubini, and Conto-type immigrant will add luster to the American firmament.*
>
> ***And when Congress does its rightful duty as urgently requested by our President and enacts the new legislation, the sad story of the U.S. loss of Dr. Gustavo Bounous will not be repeated.***

A man like Bounous, patient and steadfast though he may be, nevertheless could not wait forever. His dream of becoming an American was apparently thwarted forever by American bureaucracy. The eloquent descriptions of his plight, recounted in Congress to some of the most powerful men on earth, seemingly availed him nothing. In the end, Dr. Bounous made a rational and important personal decision: he would become a citizen of Canada. Montreal, Quebec would become his home.

"In 1967 I received Canadian citizenship," Bounous says,

adding poignantly, "for once I felt like I was someone ... a proud citizen of Canada. I actually liked being a British subject. The fact of coming from little people who endured six centuries of persecution and slaughter, and knowing that the British sympathized ...

"Oliver Cromwell helped my ancestors. Beckwith, a Scotsman, came and built churches for us. My past was aligned with Great Britain and the courageous French Hugenots as well, and these meshed and blended with my new citizenship in Quebec."

The two languages commonly spoken in his new Canadian home now gave him no trouble. He had learned to speak and read English in America, and had previously learned French in grade school.

"My religion classes were in French, since the French Hugenots had allied themselves with the Waldensians. They rallied to our cause and these two arms of Protestantism had for years served side by side."

Once decided, he entered into his new citizenship with appreciation and joy.

"I felt both proud and happy."

From Bounous's initial deep disappointment at being rejected by the United States, grew a strong attachment to his adopted country of Canada.

Canada, as it turned out, was to become not only the wanderer's haven, but also his destiny. It was in Canada that Gustavo Bounous's strong body of research developed, then gained stature and recognition. Canada financed Bounous's research investigations, provided the milieu in which a serious scientist could do serious work, and at last became the seedbed for his greatest medical discovery yet to follow.

The riveting story of how that medical breakthrough came about convinces everyone who hears it that the fickle hand of fate pointed Dr. Gustavo Bounous toward his great personal fulfillment and professional destiny. That destiny resided in Canada.

Ada Bounous

Ada and son

'Tito'

Young Dr. Bounous

Torre Pellice high school

Genevieve DeSerres & 'grandpa'

Genevieve fishing

Gustavo's home in San Giovanni, Italy

Birth place of Gustavo: Piedmont, Italy

1978

"... you are not alone, but God
is within,
and your genius"

- Epictetus

9.

RESEARCHER, EXPERIMENTALIST, ANALYST

After receiving the honor of Canada's highly coveted professional medal, Dr. Bounous was then 'discovered' by others in the medical community and soon was invited to move to the University of Sherbrooke as an Associate Professor of Surgery.

The offer very likely might not have tempted another man in Bounous's position. He was well liked at McGill, respected by his peers, and his work was going well. Sherbrooke, a small Quebec city located some two hours by car from McGill, doubtless would have appeared to most people to be far less enticing than the sprawling, swinging Montreal metropolis. Furthermore, Sherbrooke University could hardly boast of anything approaching McGill's renown. Why would a celebrated scientist, possibly at the peak of his career, choose to go in that direction?

What Sherbrooke did offer, however, was a calm climate in which a serious researcher might pursue his investigative course, away from the spotlight, the competitive atmosphere and the attendant politics. Speaking of politics, McGill University, one of the world's educational and research landmarks, stood out as a British establishment amidst all that represented Gallic Quebec. As French nationalism in Quebec intensified, Bounous, who disliked the clamor over French and English tongues, jumped at the chance to get beyond the fray. He had little tolerance for either parochialism or nationalism. He never forgot Mussolini and celebrated his freedom by choosing a quiet and productive existence. He repudiated activism in almost any form. Galileo's individualism had taken deep roots in his personality.

SHERBROOKE

It seemed a simple matter for Dr. Bounous to pile his things into his car, drive a relatively short distance, and set up at Sherbrooke the same research and classroom lecture program he had established at McGill. He could move more or less on his terms, so that he did with little apprehension.

That change of direction certainly sheds light on the true Bounous character and his independent thinking.

He was not a groupie and by no means an empire builder. He was content just to do good, if not provocative, research. His mind was free to think and so his life was open and refreshing. Only his tendency to social isolation limited his popularity. The French language itself at Sherbrooke hardly presented a problem to him; after all, it was his first language, spoken in his home, used in his religious training, and only abandoned when Mussolini mandated that Italian must be spoken throughout Italy.

Bounous's world view was more global. In his spare time, he had been a student of history and philosophy. He reveled in ideas. His personal beliefs passionately aligned with freedom of choice, a view he considered consistent with American ideals. Ever the champion of intellectual freedom, Bounous says that when he was denied access to America he mourned as one might mourn a lost love.

Much more obvious of course, and even in retrospect, is the fact that Bounous's choice of Sherbrooke hardly could rank as a brilliant career move. It was a personal move. Had he been personally ambitious he certainly would have capitalized on his McGill University exploits, using his recent acclaim as a springboard to future successes. He would succeed by default, if not by design. But that was his destiny. That he did not know.

No, it was the work itself, not just the chance of some possible future lofty position and growing public notice, that motivated Gustavo Bounous. He believed deeply in his *Elemental Diet* and its potential for helping some among the sickest of the sick return to health. He ardently desired to follow his investigative instincts regarding the role of diet, despite an almost universal disinterest within the medical community at the time.

So, by what some might term a stubborn choice, Bounous elected to labor in relative obscurity. He could live with that. In some ways, life at Sherbrooke echoed his student life and medical residency in Italy. The town's ordinariness and his lack of contact with much real life beyond the laboratory, eventually combined to create within him a familiar condition – deep-seated, pervasive loneliness.

"Most doctors and surgeons see patients, colleagues and patients' families every day," he observed. "In my lab I saw almost no one else except an undergraduate assistant or occasionally a colleague ... And then there were nuns," he added with a chuckle.

Other men living without the comforts of marriage and family might not have endured very easily a lifestyle which provided so little personal stimulus and fulfillment. Dr. Bounous, however, conditioned from boyhood to accept life with a minimum of creature comforts, was the man least likely to complain. Actually, the rather emotionally barren atmosphere proved strongly conducive to what he did best: focus with laser-keen intensity on his medical investigations. He was a true scientist in phenotype.

The Sherbrooke years, arid as they were in some respects, nevertheless produced some of the most painstakingly focused medical observations of Bounous's career. The timing, like the place, proved perfect for the missions at hand. Stuck away in a quiet corner of thriving Quebec with few others to peer over his shoulder, one of Canada's most singular medical investigators followed his nose, exploring his own various avenues of scientific questioning.

What did he seek?

"More information about what I called 'medical food'," he said. "I conducted many more experiments and clinical studies on my elemental diet."

In this, Bounous marched to the beat of a different drummer. His was at the time, the lone research voice in his field, raised to question the role of nutrition in treating not only hemorrhagic shock, but also in managing intestinal lesions caused by radiation and other deleterious insults. He was intrigued by the effects on pancreatic proteases in the intestine. He sought further uses of the elemental diet in intestinal disorder and in the critically ill.

CAREER INVESTIGATOR

Though Sherbrooke University named Dr. Bounous an Associate Professor, later elevating him to a full professorship in the Department of Surgery, in actuality he continued to work as a Career Investigator for the Medical Research Council of Canada.

"The government paid my salary," he explained. "In those days Canada supported a great deal of government-sponsored medical and scientific research. My work was one small part of that broad effort."

Bounous's vital relationship to the Medical Research Council worked like this: after careful review and diligent analysis, the doctor proposed a new project by preparing a detailed, written plan. The proposal would be funded only after a peer review and a period of consideration including perhaps debate within the Council. After receiving the green light, Bounous could proceed with his typically three-year project, working mostly alone, while keeping meticulous records and submitting interim reports. Periodically a team of three Medical Research Council - appointed physicians and/or surgeons would arrive at his office to review his work and question him at length.

These in-depth interviews were as much a stimulation as they were a provocation.

"These gentlemen were always courteous, and very knowledgeable," Bounous recalls, "but the week prior to one of those oral peer reviews, I would experience the sort of anxiety attacks I felt just before exams in medical school. I would suddenly feel that everything depended on a good outcome. In medical school, I was always anxious about my being able to support my family. In Sherbrooke, I wondered if the review board would okay my work and agree to fund the next project. That would mean my salary too. Every time I wondered if I might be about to lose my job."

Although Bounous's work would be scrupulously thorough, and despite his dependably good approval ratings, his anxieties were based in large part on solid reality. He knew very well that he too often chose the oblique approach to surgical research, continually sliding into medical avenues most other scientists chose to ignore. It was a risk he had to take, driven by his own curiosity.

When would his luck play out? At what point might he lose his funding? As it turned out, that never happened. Had his worst fears ever materialized, of course Bounous easily could have taken a position – probably a much more lucrative one – with a pharmaceutical company, for example, or found some other suitable employment. No, his fears resided within his realization that he lived on the research edge, never opting for the scientifically trendy, the question of the moment, or playing into establishment research ideas in general. He was financed by the public purse and so he could remain outside the manipulation and pressure from the establishment. (Unfortunately, public funding for medical and scientific research is on the decline - at least in Canada. Truly objective work will soon be hard to find if a few large pharmaceutical companies happen to control the research purse strings. This is good reason for public concern. Will Dr. Bounous come to represent a dying breed?)

"Many scientists pursue their investigations and complete them satisfactorily without the least interest in the *revelance* of their results," he explained. "However, I was not all that knowledgeable about any one field. I was hardly a specialist, so I could not thrive in esoterica. I did a little of everything – hemodynamics, medicine, surgery, nutrition. I had adopted a broad view of medicine and pathology which allowed me to see the relevance of my various investigations. The novelty was always at the interface of the different fields. Nothing was important in isolation."

Thus Bounous interested himself in a holistic view of the human body and its functions, while most in the mainstream of research rushed towards specialization. It would be years before many others in the world of medicine adopted such an holistic approach, the only one that seemed reasonable to him. It came naturally from his openness and honest inquiry.

Independent-thinking as always, Dr. Bounous at last was bold enough to ask one distinguished member of the Research Council's peer review team on one occasion, a question which long had plagued him.

"Off the record, doctor, does it make any sense that so many research investigations, successful as they may be, add up to nothing?"

His mentor seemed surprised, even disconcerted at the question.

"Why, no-o-o," he answered slowly. "No one else ever asked me that question. As long as it's good research, why should results matter? It's all an on-going, evolving process, isn't it?"

To Dr. Gustavo Bounous's questing, practical mind, that answer made little or no sense. Years of costly research with no expectation of any real application of the results? No usefulness? A process with never a product in view? Gustavo Bounous simply didn't get it. In some ways, he thought, the life of a research scientist would seem only whimsical and risky, against the odds. He could only think of purposefulness and accountability.

"Once, reading a biography of Beethoven, I saw where the

composer said, 'Oh my God, I wish I had a little pension,'" Bounous mused. "How I, like Beethoven, wanted a basic income! Yet I was a scientist, not a musician or artist. My world was real and tangible and needy. I must affect nature to effect change. I could only be properly paid if I did something useful."

Still, he kept going. He had to write at least two or three articles each year for publication in top-level medical journals. That was an unnecessary burden.

"At least every three years also, came this peer review, an on-site inspection of my work for two full days. The three scientists asked me countless questions and I had to provide all sorts of documentation."

At times Bounous's work earned criticism.

"They would sometimes say my work was average," he recalled. "These things keep you humble; there is absolutely no occasion for one in that sort of research position to build up any professional pride.

"How can you measure the value of your work? Alone in the laboratory with your mice and your dogs for most of the day, working only occasionally with one or two other people, you receive little or no feedback on your work. You can only exist by believing in what you are doing."

BACK TO McGILL

In 1976 the Medical Research Council suggested that Bounous might take a sabbatical year at another university to "freshen up" and gain new perspectives.

"Because I always follow the path of least resistance, I decided to return to McGill University," he said. "They still respected me and it was nearby and familiar. Besides, I hate to travel. Flying almost terrifies me. Therefore, I returned to Montreal ... by car!"

Circling back to McGill's Department of Surgery indeed

did afford Dr. Bounous a new perspective. No longer working in a professional vacuum, the very corridors and office cubicles - despite their familiar drabness - reminded him of earlier successes he had accomplished there.

The role of the intestine in hemorrhagic shock ... the intestine involves diet ... his Elemental Diet was something new ... promising ... something practical and useful ...

Yes, it really did add up, after all. His work was leading somewhere, somewhere very basic and important. He could feel it. Perhaps the Medical Research Council did not expect or require one's discoveries to have immediate applications, but Gustavo Bounous desired exactly that for himself. His hard work should produce, at some point, some really useful results ...

He was returning not only to his work but to an old familiar environment. One that he knew well and felt comfortable in. He remembered the place and quickly reestablished a routine with familiar names and faces. In the laboratory he had other friends, the animals. The small, plain office cubicles were little incubators for the researchers' active minds. The muted sounds of low telephone conversations or quiet laughter could still be heard from the tiny enclave devoted to coffee breaks, situated across the corridor to his right. Even the soft slap-slap of rubber-soled shoes against the rubber-tiled corridor floors remained. These all represented to Bounous the invigorating hum and hue within that academic beehive. It represented, in part, some of the world-class medical and surgical research at Canada's McGill.

Here, anything could happen. Some of the world's best young talent streamed through McGill's enormous portals, eager to get in. Scientific queries continually winged their way to McGill with its scientific brain trust, from the most remote areas of the world.

And something did happen ... something unexpected ...

MYSTERY IN THE MAIL

One otherwise uneventful day, the mail brought a particularly intriguing query from a large food manufacturing corporation in Switzerland to the desk of one Dr. Gustavo Bounous. It was a registered package – neatly boxed and wrapped.

The cover letter explained the reason for the $10,000 check enclosed – that was money which expert Bounous was being asked to use to explore all possible uses for the material they were sending him.

Curious, Dr. Bounous examined the box carefully. It contained a fine, white powder – some milk whey concentrate, according to the letter - a substance which was presenting the giant food company with much embarrassment and a gigantic corporate headache.

As a milk by-product from cheese manufacture, it was useless to them and thus had been routinely dumped into nearby rivers and streams. But no more. Unfortunately, the whey was rich in protein-nitrogen and it had begun to nourish local flora which threatened to clog the waterways, presenting a new major ecological challenge. Environmentalists and local government officials were up in arms. Something had to be done. They were turning to the expert.

After all, this was a food derivative. Would there be any medical or nutritional use for this unwanted whey or, at the very least, some better way to dispose of it? Dr. Bounous should know or at least, he could find out.

The Bounous reputation had gone ahead of him.

Dr. Bounous peered at the undecipherable signature at the bottom of the brief letter, wondering to whom he would respond. There was no name typed below the scrawl, no hint on the letterhead. Unconsciously, he was fingering the check with one hand. With the other, he gingerly poked a finger into the amorphous whey powder to feel its consistency, then he tested the bland stuff on the tip of his tongue. It was tasteless.

Bounous had a sudden realization. 'A——haa ...'

The corporation had sent the sample, the challenge and the money to him, *because he was now perceived as an expert on the forefront of nutrition research.*

Could that really be true?

It was.

Probably, the nameless corporate worker who decided to send this query to Dr. Gustavo Bounous held little real hope for a solution to that company's problems. Doubtless it was a long-shot effort, a call out of desperation. And for the doctor, the fine white powder on his fingertips presented a rare challenge, perhaps an opportunity, and with it, certainly a nice check. At least, to start with.

Neither the manufacturing executive in Europe nor the research scientist in Canada had the slightest idea then that the powder, metaphorically speaking, eventually might amount to *'medical dynamite'.*

By 1978 Dr. Bounous's elemental diets had demonstrably improved conditions of patients with intestinal lesions caused by shock, intestinal ischemia, radiation, cancer chemotherapy and Crohn's disease. Perhaps with different dietary manipulation, there could be other clinical influences for entirely different reasons. Dr. Bounous would not prejudge, he must remain open and let the research lead him where it would. That was the story of his research career up to that time. Now he prepared to investigate any effects of the new whey protein he had just received. Would bovine whey by some chance prove nutritionally useful?

There was only one way to find out. Dr. Bounous would feed the whey powder he received to a small group of mice and compare their response with a similar group fed a more normal maintenance diet.

The protocol was quickly established and he set about diligently to satisfy his curiosity.

He was open.

There was no qualitative difference in the general appearance, growth, attitude and activity of the mice fed with this non-specific derivative of whey. However, when challenged with T-cell dependent antigen, these mice produced more antibodies compared to the normal mice fed on the standard maintenance diets. These data were reported in 1981 in the *Journal of Infectious Diseases*.(15) Could the immune system really be affected so dramatically?

The answer seemed to be immediately self-evident.

"I saw immediately that the whey product added to the mice's diets produced a good immune-enhancing effect," Dr. Bounous recalls. Typically, however, at first Bounous refused to even believe his eyes. He reasoned that the positive results he saw must be due to some error; something fortuitous. But the 'error' seemed consistent.

COLLABORATION

Dr. Bounous was soon fortunate to be introduced to Dr. Patricia Kongshavn, one of McGill's pioneer immunologists. It was more than just coincidence that Dr. Kongshavn's investigations into the human immune system would combine well with Bounous's nutrition work. From the very meeting, the pair meshed well in their scientific interests, and the Bounous/Kongshavn teamwork continued for the next decade and beyond. Their studies and experiments were to produce startlingly significant breakthroughs in each of their respective fields.

Fortunately for Dr. Bounous, Dr. Kongshavn took an immediate interest in his project, without prejudice. An honors graduate of Cambridge University in England, with a Master's and a Ph.D. degree in immunology from McGill, Kongshavn at that time directed two laboratories funded by independent research grants from the Medical Research Council of Canada and such other sources as the National Cancer Institute of Canada and the National Institutes of Health in the United States.

"Dr. Kongshavn had the laboratory, the animals, and most importantly, an intrinsic belief in what I was doing," Dr. Bounous said. Luckily, the pair met and collaborated at exactly the right time in both their scientific careers.

"Dr. Bounous came to me with his idea and asked what was the best way to look at these possible 'immune' effects of such novel dietary change." Dr. Kongshavn recounted. "At the time I knew of no one else except Gustavo who was studying such things."

Previous research over two decades had shown that severe protein deficiency predisposes to or exacerbates bacterial infections. This obvious suspicion had been confirmed in several experimental model systems. In these studies, the test diets typically contained only 3-5% casein protein, or were devoid of protein altogether. Those diets all brought about substantial weight loss when compared to control diets containing 20-24% casein. That was not too surprising. After all, malnutrition leads to weight loss. But observations of the interaction of protein-calorie malnutrition and various immune parameters provided some understanding of the etiology or cause of the increased susceptibility to infections associated with the depleted diets. These negative effects on the immune response varied with the degree of protein deprivation.

Sure, the effect of malnutrition was easily accepted. But what about the opposite? Could changing or improving the diet enhance immune response directly?

Dr. Bounous was determined to find out.

10.

YOUR IMMUNE SYSTEM

After Dr. Bounous initially observed the effect of feeding his whey protein to laboratory mice, he became avidly interested in the immune system. He began reading on the subject more extensively than he had ever done. He had long discussions with Dr. Kongshavn and almost anyone else who was even half interested in the subject.

During one of his lectures (from among the few that he gave each year at McGill and then at Sherbrooke University) a bright precocious student asked a question about the rejection of transplanted organs. It was not truly relevant to the subject in class that day, but Dr. Bounous pounced on the question and gave a five or six minute answer, much to the satisfaction of his students. They were obviously interested in the subject and a few students even hung around in the hallway outside after class to query the specifics further. Dr. Bounous was so animated by the questions that he went on to share with these eager undergraduate students about his new project and the challenges it posed.

The few students were fascinated. They wanted to know more. They were not yet fully prepared with the requisite biology course in which they would later receive a proper introduction to Immunology. They were therefore not to be ashamed at all for their apparent ignorance of the same. That knowledge would come later.

Dr. Bounous, seizing the moment like a true academic, arranged for a special tutoring session to follow up this encounter. He invited them to come by his office and he would prepare a briefing on the immune system for them. Perhaps the reader will find some value in this brief overview of the immune system too.

OVERVIEW

For your constant protection, your body is equipped with a number of anatomical, physiological and biochemical systems that work in harmony to defend your life and health. The principal components are summarized in Table 1. Following are some key characteristics and what you can do about them.

THE SKIN

The skin is the largest organ in the human body. It performs many crucial functions including being an effective barrier to environmental threats, both physical and biological. When anyone suffers from severe burns for example, there is a medical crisis due to the extreme challenges to the body's defenses.

KEY: *To avoid dehydration and infection. Avoid sun damage. Treat burns.*

THE RESPIRATORY TRACT

Special linings in the airways, including the nasal passages and throat, protect the lungs from foreign particulate matter and also from smoke and chemical inhalation. The latter usually triggers coughing to facilitate expulsion.

KEY: *To avoid polluted air and cigarette smoke.*

TABLE 1. OVERVIEW OF YOUR BODY'S PROTECTION

EXTERNAL		
- THE SKIN	➪	Barrier
- EPITHELIAL LININGS		
* Respiratory Tract	➪	Breathing
* Gastrointestinal	➪	Eating
* Urogenital Tract	➪	Intimacy
* Auditory Canal	➪	Hearing
INTERNAL		
- ORGANS		
* Liver	➪	Pollution Plant
* Kidneys	➪	Filtration
* Lymph Nodes & Spleen	➪	Processing
* Bone Marrow	➪	Production
* Thymus Gland	➪	Maturation
- CELLS		
* The Immune System	➪	Defense Force
- MOLECULES		
* Antibodies	➪	Weapons
* Cytokines	➪	Triggers
* Antioxidants	➪	Stabilizers

THE GASTRO-INTESTINAL TRACT

The daily intake of food and water passes through almost forty feet of complex tubing during which digestion and absorption are critically controlled, physiologically. Vomiting and diarrhea are protective mechanisms of emptying the bowel. Chemical secretions and special linings defend against harmful contents which remain.

KEY: *To exclude ingested poisons and metabolites. Eat healthy foods.*

THE LIVER

The central processing of waste that gets into (or is formed in) body tissues, takes place in the liver. Many complex enzyme systems there, convert harmful products into harmless waste that is secreted into the blood stream and cleared through the kidneys or the biliary tree (urine and bile). Advanced liver failure could allow toxic products to get to the brain and usually leads to a medical emergency.

KEY: *To avoid addictions, additives and substance abuse.*

THE BONE MARROW

The bone marrow is a very productive organ. All blood cell lines are produced in the bone marrow, including the cells of the immune system. Today, bone marrow transplantation is being used in life-saving procedures to restore blood cell production in certain clinical states.

KEY: *A healthy, complete diet. Moderate exercise.*

THE LYMPH NODES

Almost everyone at some time has experienced swollen lymph glands or nodes. These are usually indicative of infection, or sometimes reactive inflammation and even cancer. They represent small factories for immune cells and their defensive activities. Some cells produce antibodies that are effective circulating weapons which trigger a cascade of chemical and biological warfare. Hopefully, the immune system wins.

KEY: *Monitor and report.*

IMPORTANCE

Everyday and in every way ... your body is always under attack, but your immune system protects you both in health and disease.

Just imagine, your health is indeed constantly threatened from bacteria, viruses, radiation, chemicals from without and within, as well as from mutations, trauma, etc. In fact, when one reflects on the wide variety of insults to which each of us is exposed every day, it is surprising that we do not fall victim to illness and disease more frequently. No person is an island and we do not live in a controlled biosphere. On the contrary, we are surrounded by a world of hostile microorganisms that do all they can to survive at our expense, and an environment of chemicals and radiation that can wipe us out, if we are not careful.

Therefore, the importance of the immune system cannot be overemphasized. When all the professional health care workers have done their best, one must still be obliged to wait and see how any given person will react in the face of whatever threatening health situation they may encounter. In a sense, that is the ultimate measure of true physical fitness.

When exposed to insult or challenge, your cells fight back by a complex series of *inter-* and *intra-* cellular events, which in total, is generally referred to as the immune response. They declare war on the invading enemy. They could lose a battle, now and then, but they must win the war! In real terms, your health therefore depends on you or more precisely, on your immune system. In fact, your life depends on it!

For the past few decades, scientists have actively studied the body's immune system. They have made great strides in understanding how this complex system works. Such research will continue to have increasingly widespread implications for both health and disease. But you need not wait, you can do something about it now.

CHALLENGES

The immune system has at least four broad challenges to reckon with:

(i) *To identify the enemy.* The system must be able to distinguish what is self or non-self, friend or foe, for you or against you.

(ii) *To carry out immuno-surveillance.* It is important that the response be specific but it must also be on alert so that its defense is quick and overwhelming before the enemy strikes a serious blow.

(iii) *To fight back.* It must be effective and complete in devastating the challenge from any perceived enemy. This is war and not a cellular picnic. It must go for the kill.

(iv) *To turn-off or shut-down when it has done its job.* The immune response is truly one of defense. It must never be overzealous so as to be offensive against the body itself. There must be controls.

All these challenges must be met sucessfully and simultaneously. To that end, the immune system has a highly sophisticated repertoire of specialized cells which execute their varied functions in a highly coordinated manner. The entire system is internally regulated by genetic coding of the differentrated cell lines. Only the cells of the immune system are sufficient for the task. Clinical intervention at best, only seeks to exploit what nature does efficiently and consistently by cellular design.

RESPONSES

In the most general terms, nearly all insults to your body in both health and disease, will result in two major consequences: First, there is a general physiological response from your organs, acting in concert to discern different priorities and re-distribute the bodies resources. Then there is the cellular response which gets right down to defense at the microscopic level.

1. The GENERAL Response or Stress Adaptation Syndrome

This is a common pattern of retaliation, by which your body prepares for fight or flight. Most of this happens spontaneously by reflex action.

At the MACRO level: The major physiological systems adapt under varying conditions to maximize your chances of survival and safeguard your health. For example, you can experience your heart racing, fever, exposed blood clotting, changes in urinary output, circulation, gut motility, muscle tone... etc.

The net effect is to maintain or restore what is called the state of 'homeostasis' in which the healthy body thrives. It was the famous 19th century French physiologist, Claude Bernard, who first observed that "*the constancy of the internal environment (i.e. homeostasis) is the condition for a free life*".

Life is maintained then in a healthy state only within very narrow ranges of all the physiological variables. Feedback and reflex control mechanisms allow the body to adapt quickly and effectively when challenged. Survival itself may be at stake but at least, our state of health is critically determined by this adaptation capacity. This is the most important sign of robust or dynamic health.

2. The CELLULAR and Immune Response.

As was noted before, this second major consequence arising from any insult to the body, is targeted at the local level. It may take different forms, but the overall strategy may be classified broadly into an inflammatory response and an immune response.

The *inflammatory* response is somewhat more pathological but essentially, it is a process consisting of a dynamic complex of reactions that occur in the affected blood vessels and related tissues, in response to any injury or abnormal stimulation caused by some physical, chemical or biological agent. The classical cardinal signs are *rubor* (redness), *calor* (heat), *tumor*(swelling) and *dolor* (pain). Function can also be affected.

Inflammatory changes are therefore usually gross enough to be observed. They occur at the tissue level even though they are triggered by complex cellular activity.

The *immune* response on the other hand, is generally invisible though no less significant. As we have said, it is cellular warfare. The cell is under attack and it is prepared to fight back.

At the MICRO level: Each cell is a living entity that has unique characteristics of 'homeostasis' too, even in the microscopic world. The cell exploits molecular biology and genetics to first protect itself. Then the specific cells of the immune system are able to participate in an orchestrated collective response that is differentiated and very specific. This spontaneous retaliation is most effective in defense of organic life and is subject to regulatory control.

There are 3 major challenges that confront **all** your cells *internally*:

* **F**ree radicals ('hot' molecular fragments)
* **O**xidation (oxygenated products)
* **X**enobiotics (poisons)

Free Radicals

Chemical bonds between atoms typically involve the sharing of paired electrons. This leads to stable covalent molecules. When a chemical bond is broken by some reaction or another, the fragments produced each have 'unpaired' electrons which make them most unstable and therefore highly reactive. They become chemically 'hot' and attack other susceptible molecules in search of electrons. These may include critical nucleic acids or proteins and when DNA or RNA is involved, genetic control of cellular activity is compromised. ***Free Radicals must therefore be quenched before they damage the tissues.***

Oxidation

Most important chemical reactions inside cells often (in fact, usually) involve oxygen or other reduction – oxidation systems. The simple transfer of an electron or proton from one atom to another has major consequences. Oxygen is the ubiquitous key and must be controlled. When premature oxidation takes place in cells with superoxide-, hydroxyl- and peroxy- radicals as the lethal intermediates, cells can be devastated. ***These toxic intermediates must be neutralized before they cause a cascade to cell death.***

Xenobiotics

Poisonous chemicals can be inhaled or ingested and they eventually would end up in the circulation and body tissues. Other products of digestion and metabolism can also become toxic to cells. ***All these must be neutralized before they cause damage.*** Those formed inside the cell are best neutralized on-site inside the cell. Other xenobiotics can be transported to the liver to be neutralized and transformed into more harmless products and later excreted via the kidneys and biliary tract.

YOUR DEFENSE FORCE

ANTIGEN. This could be any macromolecule which, on being recognized by the immune system as **foreign**, stimulates an immune response. That response is specific.

ANTIBODY. This is one of a wide variety of secreted proteins elicited in an animal by different antigens, which reacts specifically with each particular antigen to **trigger** a variety of outcomes. They can slow down and kill viruses and bacteria.

PRIMARY RESPONSE. The formation of antibody after **first** exposure to an antigen.

SECONDARY RESPONSE. The formation of antibody after a **second** exposure to an antigen. It is an accelerated response (that is, faster) and an exaggerated response (that is, of much greater magnitude).

T – CELLS. A variety of lymphocytes originate in the bone marrow and are processed by the **Thymus gland**. They have specific receptors. In general, T-cells of one type have effector or regulator functions, while another type are "killer cells."

B – CELLS. These lymphocytes are mainly generated in the **Bone marrow.** They are defined by the expression of immunoglobulins on their surface which form specific receptors. When stimulated, they proliferate, mature into plasma cells and secrete antibody.

Both T - Cell and B - Cell lymphocytes have specific receptors which combine with foreign material of complementary structure and undergo clonal expansion by the millions. That is a unique characteristic of the immune system.

ANTIGEN PROCESSING CELLS. There are several morphologically distinct classes of cells which process antigens non-specifically and prepare them for **presentation** to the T-helper cells.

T – HELPER CELLS. They are small lymphocytes but act as the **Commanders-in-Chief** of the immune system. They recognize processed antigen, become activated, and in turn, trigger responses from other members of the defense force. They are distinguished by CD4 markers.

T–KILLER CELLS. These **Paratroopers** activated by the T-helper cells have only one goal: to destroy the enemy before it has time to multiply. They wipe-out virus-infected cells and perform immune surveillance. They are distinguished by CD8 markers.

B – CELLS. Activated specifically again by T-helper cells, these specialize in **chemical warfare**. They then mature and produce immunoglobulins and secrete antibodies which circulate in body fluids to find antigens and trigger immune responses (usually, that means death to the enemy).

K – CELLS. These are the immune **Thugs**. They are subpopulations of non-B, non-T mononuclear cells and are responsible for lysis (destruction) of cells with antigen - antibody complexes.

NK – CELLS. These are medium sized lymphocytes more like **Hired Thugs**. They also have no B-cell or T-cell markers but possess characteristic granules. They can, without specific recognition, spontaneously kill-off transformed (cancerous) cells or virally infected cells.

The immune response is carefully orchestrated. Each type of cell is genetically programmed and equipped for its specialized tasks.

MACROPHAGES. They are the **Ground Forces** that roam throughout the body, consuming pollutants and other invaders at a restless rate. Their activity can be enhanced by antigen-specific T-cells which release soluble products. When activated in this manner, macrophages can distinguish between normal and transformed cells.

MAST CELLS. These are the **Weathermen**. Large cells (and their precursor, basophils) bind specifically with a class of immuno-globulins which then combine with antigens in bridge-pairs resulting in release of vasoactive amines, particularly **histamine**. This histamine release is responsible for the clinical allergy reactions.

T – SUPPRESSOR CELLS. They are the **Flag-Bearers**. Small lymphocytes, they will call off the battle when victory has been achieved. They tell T-KILLER cells to stop the fight. They also carry the CD8 marker.

T – MEMORY CELLS. These cells stay in the body after the primary exposure or infection. They are **Data-Processors** trained to recognize the invasion of the same enemy in the future.

B – MEMORY CELLS. They function in the same way as T-memory cells. They will produce antibodies in the secondary response, upon recognizing the same enemy. They initiate rapid, exaggerated response like **Marines**. That's the basis for immunization vaccines.

The Immune Response is usually protective but if the system fails to be discriminating, civil war erupts. The body can then treat particular normal cells or parts as the enemy – that's the basis of many auto-immune diseases.

YOU HAVE A DEFENSE SYSTEM THAT IS SPECIFIC, EFFICIENT AND INDISPENSIBLE.

The McGill students were most appreciative of those notes. They gained a small insight into this new, exciting area of biological science. Like any good teacher, Dr. Bounous was delighted to plant those seeds. But he was even more delighted to return to his laboratory where he was increasingly fascinated by the new discoveries he was making.

11.

DISCOVERIES !

Dr. Bounous began his whey protein studies in earnest during his sabbatical year back at McGill. Working closely with Dr. Kongshavn the immunologist, he was able to accelerate his learning curve in this new field of interest. Their combined interest focused initially on the effect of dietary amino acids (protein building blocks) on immune reactivity. Later they would look at a variety of common dietary proteins. To measure the immune response, Kongshavn suggested they use two fairly standard methods.

i) The Plaque Forming Cell (PFC) assay for *humoral immune responses* (antibody production). This method was used for assaying the immune response, as modified by Cunningham and Szenberg. (15a) Mice were injected intravenously (i.v.) with sheep red-blood-cells (sRBC) and then the spleen was assayed for plaque-forming-cells (PFC) five days after inoculation when the response was shown to peak.

ii) Mitogen responses. The method described by Lapp and co-workers (15b) was used to test the mitogen response to different concentrations of phytohemagglutinin (PHA), concanavalin A (con A) and lipopolysaccharide (LPS) mitogens in the spleen, with or without stimulation with BCG mycobacterium. The PHA and Con A responses were indicative of the other fundamental component of the immune system i.e. the *cellular immune response*, while the LPS response measured the B cell responses.

The first results of Bounous and Kongshavn were more definitive for the negative effect of severe dietary restriction, particularly of some essential amino acids. They tentatively proposed that this caused a suppression of the production or function of some inhibitory cell such as a T-suppressor cell, while not affecting the influencing cell to the same degree.

But the question still remained. Was there some special dietary manipulation that could *positively* and *consistently* enhance the immune response? Bounous and Kongshavn were in pursuit of an answer.

IMMUNO-ENHANCEMENT

The pair turned to investigating various edible dietary proteins – including casein, lactalbumin or whey protein of milk, soy and wheat. Of these, lactalbumin uniquely enabled the mice to develop an immune response which was consistently greater than that of comparable mice which were fed the other protein diets.

"It was distinctly different," she was quick to point out. "As we continued to do research and include other dietary proteins, we were surprised to find that, of all the ones we tried, only one in fact strongly enhanced the immune response – only the whey protein concentrate (lactalbumin) had this effect and, at first at least ... it did so every time."

Dr. Kongshavn terms those first results interesting and publishable(16, 17), but not yet really exciting.

"We continued to collaborate," she says, "and I certainly continued to be interested in Dr. Bounous's belief that the immune system could be altered by diet – a very original idea."

When his sabbatical year at McGill ended, Dr. Bounous returned to his post at Sherbrooke. He kept his apartment in Montreal so he could commute and continue working with Dr. Kongshavn.

"We maintained a rather loose contact," she recalls, "yet always worked together closely."

With Bounous commuting and Kongshavn running her own research laboratories, both doctors also lecturing, supervising technicians and graduate students, and writing the indispensable scientific papers for publication in reputable journals, those became some of the most relentlessly busy years of their respective careers.

"As a Professor in Canada you are essentially your own boss which is immensely satisfying but, on the negative side, this means that you are also responsible for raising your own funding for research," Kongshavn reminded me. "Gustavo was working in a very non-fashionable area. He was working alone. He'd come to Montreal a couple of days a week and we'd look at results and decide what to do next, or we'd work together on papers or write grant proposals."

Though the two scientists worked in tandem, Kongshavn says Bounous always decided on the diets.

"You have to design your experiment in a way that allows you to see the results," Kongshavn explained. "That's where my expertise in Immunology came in. When one is looking for a particular effect, it may be there but masked by other factors. In order to 'see it', therefore, it is necessary to select exactly the right set of conditions. For example, had we given our mice the 'standard' immensely high doses of antigen usually used, the immune response could have been so great that the more subtle differences we observed between mice fed the different diets could have been missed."

Mice are the laboratory animals of choice in immunological research. They have afforded major advances in the understanding of the immune system, which has led to many applications in clinical medicine. They are mammals and therefore extrapolation can be made for application to humans. There is also a vast body of reference data for consistent comparison. They are available to scientists in select genetically identical strains, with defined phenotypic characteristics.

In a typical Bounous experiment, the mice would be purchased from the breeders at five or six weeks of age. They were housed in wire-bottomed cages to prevent coprophagy (i.e. to maintain a clean hygienic environment). They were fed by placing feedings (typically three times a week) in stainless-steel dispensers for continuous availability of the powder without spillage or contamination. Drinking water was usually allowed *ad libitum* (freely). Various diets would be commenced at six to eight weeks of age and immunological studies initiated one, two or three weeks later. The mice were commonly fed in different but comparable dietary groups of ten or twelve each.

The effect of graded amounts of dietary lactalbumin (whey protein concentrate) (L) and casein (C) hydrolyzates on the immune responsiveness of two different strains of mice was investigated by measuring both the specific humoral immune response to sheep red blood cells (sRBC) and the nonspecific splenic cell responsiveness to mitogens after BCG stimulation. The nutritional efficiency of these diets was similar at both 12 and 28% amino acid levels. The immune responses of mice fed the L diets were found to be significantly greater than those of mice fed the corresponding C diets, especially at the 28% level. Furthermore in mice fed the L diet, increasing the concentration of amino acid in the diet from 12 to 28% greatly enhanced immune responsiveness by both parameters measured. In the C-fed mice, a comparable enhancement of mitogen responsiveness with increasing amino acid level of diet was seen, but there was no change in the humoral immune response. These dietary effects on immune responsiveness were remarkably similar in both mice strains tested.

The duo always used mice which were genetically identical and especially bred for research purposes. They conducted test after test, with consistent results: mice fed milk whey concentrate (specifically 'lactalbumin hydrolyzate') fared significantly better than mice given standard commercial chow or casein protein hydrolyzate.

"Lactalbumin" was the term traditionally used to describe the *group* of milk proteins that remain soluble in "milk serum" or whey after precipitation of casein at pH 4.6 and 20°C, as in the normal manufacture of cheese. The major components of the whey protein mixture were actually determined to be beta-lactoglobulin, alpha-lactalbumin, serum albumin and immunoglobulin.

Bounous and Kongshavn demonstrated unequivocally that mice fed diets containing any one of the major *components* of "lactalbumin," currently called "whey protein concentrate", or "isolate", in the concentration of 20% by weight in the diet, developed immune responses to sheep red blood cells which were inferior to that of mice fed a diet containing 20% by weight of the mixture (lactalbumin). Hence, the assumption was made that the immunoenhancing effect of lactalbumin was dependent on the overall amino acid pattern resulting from the contribution of all its protein components(18).

There was no equivalent substitute: not free amino acids, neither casein, soy, wheat or corn protein, egg albumin, beef or fish protein, *Spirulina maxima* or *Scenedesmus* algae protein, or Purina mouse chow. These were all tried and they all failed. Only with the whey powder could the immunoenhancing effect be manifest after two weeks and persist for at least eight weeks of dietary treatment. Mixing lactalbumin with casein or soy protein in a 20% diet formula significantly enhanced the immune response in comparison to that of mice which were fed diets containing either 20% soy protein or casein. Obviously whey somehow enhanced immune response, both Bounous and Kongshavn now believed ... but how? ... How?

"What we were interested in," Kongshavn said, "was an intriguing question. Where was this effect happening? Was it in the bone marrow? Or the genesis of the B-lymphocyte? If we even knew 'where or what' at least, we could then begin to think of 'how or why'. It worked, but we didn't know any more than that. What exactly was happening? That was still a mystery. We had no idea."

INCREASED RESISTANCE

While some experiments lasted two or three months, such as measuring B-cell genesis by microscopic enumeration and evaluations, most took about four weeks.

"We'd see results at that time," Kongshavn recalled, "with measurable changes as early as two or three weeks. It was never boring."

To excite things a bit more, Bounous and Kongshavn took their observations a step further. Surely if the immune system was enhanced by feeding the mice with this whey protein concentrate, hydrolyzate or intact, then one might expect these same mice to have increased resistance to some common pathogens. This they could test fairly easily. They would have to expose the little creatures to bugs and hope for their protection with the special diet of whey.

Very early in their efforts, the pair decided to inoculate the animals orally with *Salmonella typhimurium* and later with *Streptococcus pneumoniae, type 3* and *Escherichia coli,* Kongshavn recalled. Dr. Bounous, alone in the laboratory except for the technician who took care of the lab and the animals, evaluated the results. The mice, though virtually identical before innoculation, had widely varying fates: some died, others became sick, but those fed milk whey concentrate appeared brighter and healthy as they scampered about in their cage.

Dr. Kongshavn remembers those days and her surprise one afternoon when she answered the telephone to hear an unusually excited Dr. Bounous on the line.

"It works! It really works!", he shouted. He could hardly contain himself. Together, they shared their enthusiasm.

"We had injected the mice with bacteria that cause disease – real, live infection – and those mice which were fed the whey protein concentrate had survived dramatically better," she explained.

Obviously the whey powder worked (19) ... but the two research scientists still did not know why. Why would a whey protein concentrate have any such immuno-enhancing effect? What special property might this powder have? They were intrigued and challenged.

In a follow up study(20), Bounous and Kongshavn collaborated with Dr. Osmond, another colleague at McGill, to examine the genesis of B lymphocytes in the bone marrow, using two different strains of mice. The findings indicated that the observed effects of altered dietary protein type on humoral immune responsiveness were not exerted centrally on the rate of primary B-lymphocyte production in the bone marrow, but more likely, reflected changes either in the functional responsiveness of the B-lymphocytes themselves or in the processes leading to their activation and differentiation in the peripheral lymphoid tissues.

Because minerals and trace metals including zinc and copper had been found to influence the immune response, it was necessary to eliminate the possibility that the dietary proteins were influencing their rate of absorption or bioavailability. That evidence was clear. A previous study had also shown that the principal factor responsible for the observed differential effect of dietary lactalbumin and casein on humoral immunity was not the availability or concentration of single amino acids but rather the composite effect of the specific amino acid distribution in the protein.

Later, the *immodulatory effects of dietary whey proteins in mice* were confirmed independently by Wang and Watson at the CSIRO Division of Animal Health in Australia(21). Again they showed that ingestion of bovine milk whey proteins, either as a supplement in an adequately balanced commercial diet or as the only protein source in a balanced diet, consistently enhanced secondary humoral antibody responses when compared with other protein sources such as soybean protein isolate and ovine colostral whey proteins. The effects were again unrelated solely to the nutritional factors.

Such independent confirmation is always to be expected in the reporting of good science.

An adequate intake of essential amino acids is necessary because surplus amino acids are not stored and, for protein synthesis to proceed, all the indispensable amino acids must be present simultaneously in the extracellular pool. With regard to the humoral immune response, clonal expansion and antibody production require rapid protein synthesis, so that amino acid restriction will inevitably interfere with these functions. *But why is the 'distribution' and not the 'adequacy' the point at issue with these mice diets*? Questions like this not only haunted Drs. Bounous and Kongshavn but it stimulated their creative minds. One experiment led to the other. Every result prompted a different question and every question generated a new idea for experimental design. That's the normal process of science.

ANTI-TUMOR

Bounous and Kongshavn continued to measure immune responses by observing antibody responses of normal mice to different antigens they introduced into the animals' systems. They used sheep and horse antigens routinely, for example, and later, they looked at chemically-induced colon cancer cells, in which they found the mice fed with whey produced far fewer tumors than expected …There was that big 'C' word – Cancer! Cancer cells. There's the big one. Whey concentrate, they also observed, actually caused tumors in another group of mice to shrink in another classic experiment.

The cancer studies actually began with a study designed to evaluate the effect of dietary amino acid on the growth of tumors (22, 23). So much attention had been focused in the literature on the effect of food on cancer development but that was dominated by the role of dietary fiber and fat. It had become very clear that fiber was good and fat was bad. But not much was known about the influence of protein intake on carcinogenesis. The few studies in this area had concentrated on the quantity of protein intake and its amino acid supply rather than its source (24). Only limited data was available on the effect of protein type in nutritionally adequate and similar diets on the development of tumors.

Dr. Bounous and his colleagues therefore set out to determine the effect of whey protein in diets on the development of a chemically-induced type of murine tumor. In the earlier studies, the immunoenhancing property was found to be maximized at a 20% concentration of whey protein in the diet. That meant 20gm whey protein per 100gm diet. In fact, they found that raising the protein level of either whey protein itself, or casein, soy or wheat protein in the diet above 20% failed to enhance the immune response of the host beyond the values observed with the 20gm protein per 100gm diet. In addition, at this level, most proteins including those used in the test formula diets, supplies the minimum requirement of all indispensable amino acids for the growing mouse. So they chose a 20% protein concentration for the cancer study.

1,2 – Dimethylhydrazine had been demonstrated by Rogers and Nauss and by Alinen to be a convenient potent carcinogen which produced rodent carcinomas of the colon in a reproducible manner. In other words, it could be used as an animal model of colon cancer relevant to human disease (25). Others had previously shown that fiber, fat and the level of dietary protein could be either protective (fiber) or promotive (fat, protein) in dimethylhydrazine – induced colon carcinogenesis. The tumors were characteristically located in the distal bowel and long term exposure to the carcinogen was required before the lesions appeared. (Parenthetically, colon cancer in mice is also a convenient model because the response to chemotherapy is similar to human cancers.)

Dr. Bounous therefore chose dimethylhydrazine and thirty mice of a specific strain, for the classic study of the effect of his new amazing, immunoenhancing whey protein concentrate. He chose to divide the mice into three equal groups of ten which were individually numbered and housed in similar cages with five animals per cage. The mice were obtained at six to-eight weeks of age and then started on the test diets three weeks prior to commencing carcinogen treatment.

Three test diets were prepared with 20% of either whey protein concentrate or casein, or the usual Purina mouse chow (estimated 23% protein). The only variable in the two purified diets was the type of protein. They also included 56% of a protein-free diet powder containing corn syrup, corn oil, tapioca starch, vitamins and minerals, 18% cornstarch, 2% wheat fiber, 0.5% vit-iron premix, 2.65% potassium chloride and 0.84% sodium chloride (salt).

The stage was set.

The carcinogen was prepared by dissolving the powdered chemical in normal saline to a concentration of 15mg/100ml with the pH adjusted to 6.9-7.0 using saturated sodium hydroxide. Carcinogen solutions were used on the same day as they were prepared. The mice were fed the different diets for three weeks initiation and the test diets were maintained throughout the duration of the experiment. They were injected subcutaneously with a weekly dose of 15mg dimethylhydrazine per kilogram of body weight for twenty four weeks.

The animals were sacrificed four weeks after their 24th injection. Their colons were removed, opened longitudinally, fecal contents removed, and the colons then weighed and their length measured. Tumor burden was assessed both by the number of tumors and the sum of the vertical and horizontal tumor diameters of all grossly visible tumors.

'Grossly visible tumors'? Yes, one could usually see the lesions with the naked eye. And they were moreover, gross to look at.

Well, what did Dr. Bounous find on that memorable morning when he got to the laboratory to do the dissection just outlined above?

What did he see when he opened the three test groups of mice? The whey-protein fed animals developed significantly fewer tumors per animal and moreover, the tumor area development was also significantly less. This was seen despite the similarity in body weight curves, apparently ruling out conventional nutritional factors to account for the observed differences in the development of tumors.

It was well known at that time that the incidence and size of tumors were influenced by the immune system. In the advanced state of disease in which Bounous also made plaque-forming-cell (PFC) measurements, the humoral immune response was greatly reduced in all three dietary groups in the classic cancer study. The measurements, nevertheless, reflected the pattern of humoral immune response in relation to food protein type that they had earlier seen in their laboratory on healthy mice. It was therefore conceivable that, particularly in the early phase of tumor development, the protein-related differences in immune reactivity among the three dietary groups could have influenced the observed difference in tumor development that Dr. Bounous had just seen with his very eyes.

Dr. Bounous went on to publish these amazing observations in a classic paper published in the journal of *Clinical and Investigative Medicine* (1988). He coauthored it with several of his colleagues: They were bold enough to give the paper (26) its provocative title claim: "**Dietary whey protein inhibits the development of dimethylhydrazine-induced malignancy**."

It was now made public for the establishment to see.

These results would be later confirmed in *rats* by a different group of Australian investigators (27). Again, such confirmation is the hallmark of good research.

ANTI - AGING

Equally significant, the doctors became convinced that the whey protein concentrate may possess both prophylactic and therapeutic value (28). In these early classic studies, Bounous and his team had also noted coincidentally that the whey-fed mice not only had an enhanced immune response and an apparent anti-tumor activity, but in general they fared better and showed better survival rates. What did this mean in effect? Was it as basic an observation as an anti-aging phenomenon? Those were provocative questions.

Studies performed at the Eppley Cancer Center in Nebraska(29,30) were consistent with Bounous's earlier findings on the immunoenhancing effect of dietary lactalbumin. Survival (resistance to spontaneous diseases) of *hamsters* of both sexes, measured over a twenty week period of feeding from four weeks of age, was best with a 20% lactalbumin diet, in comparison with a 20% methionine and cysteine supplemented casein diet. Body weight gains were similar in both groups to suggest overall nutritional protein equivalency. In addition too, in lifetime feeding studies, the mean and maximal longevity of female and male hamsters fed 10, 20 and 40% lactalbumin diets was increased in comparison with those fed commercial laboratory feed with estimated 24% protein from various sources. Survival was again best with the 20% lactalbumin diet. In the males, longevity increased by 50%. No relationship was noted between food intake, maximal weight, and longevity.

Again, the lactalbumin diet increased survival and longevity beyond that of 'control' animals fed either of two nutritionally adequate reference diets, thus enhancing life expectancy beyond the limits traditionally assumed to be "normal". Bounous found that in one group of mice whey protein concentrate actually increased their life spans by thirty percent over three months (31).

'MEDICAL DYNAMITE'

Could this then all add up to a general anti-aging effect of Bounous's whey protein concentrate? Was it truly protective and life-sustaining at least for these experimental animals?

First, Bounous and coworkers found a surprising immunoenhancing effect in the mice, then an anti-tumor effect and now a possible anti-aging observation. Could they be handling 'medical dynamite' indeed? The answers were not yet clear.

The earliest data of all long term feeding done in the Bounous laboratory were indeed consistent with the concept that the immuno-enhancing effect of dietary lactalbumin was not a short-lived phenomenon. It could indeed be experienced as long as the protein was ingested. Furthermore, these related observations now seen in two different mammalian species (hamsters and mice) suggested the possibility that the intake of lactalbumin could, in principle, produce similar effects in man. That was a bold extrapolation then, a decade ago, and no clinical trials had been reported as yet anywhere on that subject. However, since the time of Hippocrates and throughout the Middle Ages, whey had been prescribed for human patients in large doses (up to 2 litres/day, equivalent to 10 gm/day of whey proteins) in the treatment of numerous ailments, especially acute septic conditions. Was there a correlation?

This concept may have been preserved in two old proverbs reported to originate from the region of Florence, Italy:

i) if you want to live a healthy and active life, drink whey and dine early
(*chi vuol viver sano e lesto beve scotta e cena presto*),

ii) if everyone were raised on whey, doctors would be bankrupt
(*allevato con la scotta il dottore e in bancarotta*)

These traditional proverbs were now taking on a new meaning, thanks to the early classic studies from the Bounous laboratory. Amazing observations had been made but the basic understanding was still missing. Many questions remained.

Dr. Kongshavn, who ran her own laboratory as well as the one she shared with Dr. Bounous at that time, kept insisting that the whey project was primarily Dr. Bounous's work.

"My work was primarily concerned with investigating the complex nature and function of the immune system itself – how and why the various cells interacted to bring about an immune response to a foreign invader," she explained. "So it was immensely satisfying for me to see his dramatic laboratory results that undeniably had an immediate *practical* application – a relatively simple way to improve the performance of the immune system!"

The dietary investigations, still very much on the outskirts of usual scientific interest, continued to occupy Bounous and Kongshavn as the eighties rolled on. By 1986 he had completely moved back to McGill University and a full-time research regimen there. He or Kongshavn, and often both scientists together, devoted long hours and sometimes weekends to their work.

In 1981 their report, "*The Effect of Dietary Amino Acid on the Growth of Tumors*"(32) had hardly caused a ripple in medicine's collective consciousness. The nutritional substrate was too ordinary. It held no pharmacological appeal.

"Even today, medicine still has not widely accepted the fact that nutrition undeniably plays an enormous role in both prevention and treatment of major diseases," Dr. Bounous commented. "During the eighties, we definitely stood on the research fringes!"

The pair persisted, conducting unceasing investigations into the role of whey protein in building the immune system, meanwhile exploring every avenue which might lead them to understanding how and why whey protein was so effective. The scientific papers piled up; they wrote and published at least two or three each year.

With relentless research investigations and publications in such prestigious professional journals as *"Gastroenterology"*, *"Clinical and Investigative Medicine"*, *"Journal of Nutrition"* *"Immunology"* and *"Journal of Infectious Diseases"*, Dr. Bounous, with Dr.

Kongshavn and a few others in their fields, continued to make advances by observing the amazing effects of this "lactalbumin," but the fundamentals still remained obscure.

SURPRISE, SURPRISE

Bounous, of course, had become convinced of the effects of nutrition on the immune system. He had immersed himself in intense efforts to explain those obvious effects that he was observing. Throughout, he maintained his discipline. He made steady efforts, with careful observations, while meticulous record-keeping and non-stop thinking defined his days. *Cause and effect ...? what if ...? why not try ...? Could there be ...? What mechanism ...?* Questions like these haunted his thoughts. He became obsessed, frustrated, curious, confused, determined ... all at the same time.

In his mid fifties now, well-respected by his colleagues and well-established as a serious research scientist, even if some believed he was devoting himself to less-important projects, Bounous hurled himself daily against the unyielding rock of a large and little-known subject: the medical implications of basic nutrition.

Swimming against the current would exhaust any of us. One might expect that the steady buildup of their novel observations and the convincing evidence might reward competent researchers like Bounous, Kongshavn and perhaps a few of their other colleagues with some professional satisfaction. They had got publishable results for sure. But on the broader medical scene, how many really cared? Even Bounous's general optimism and sense of cheerful expectation could lessen occasionally, especially on a long day when the troublesome hip pain he noticed as a boy chose to reappear. And it did.

"Good thing I didn't get to practice surgery," he told himself. "Long hours of standing would have made this situation even worse. I was Board-certified in Canada as a surgeon but I never practiced in the O.R. as such. I knew my physical limitations."

At the end of a long day at the lab he'd come home to relax in the evenings before the television set. He still enjoyed cartoons. They were so simple, uncomplicated, funny and mostly non-violent – perfect antidotes to challenging real-world problems and a good complement to good science.

Even in his laboratory, as demanding as his experiments became, all was not hard work, frustration and tedium, however. Despite all his efforts to maintain a professional detachment about his work, at times high excitement broke through Bounous's trained objectivity. He could become emotional. In fact, he even did admit it. In addition to the morning he first saw the denuded intestinal mucosa with hemorrhagic shock (recall the Year of the Dog), there is one other special day that he recalls even now with emotion and pride.

"It happened most spectacularly when we studied the incidence of chemically induced tumors in the mice. We had little idea what to expect. Normally we would get tumors in the colon which would spread to the liver. But the big question was: What would happen to the mice we fed the whey protein? As usual, I used thirty genetically similar mice and divided them into three equivalent groups. Again as usual, we fed one group commercial chow, adding casein to another group's ration, and whey protein to that of the third group. After several weeks I operated, with my technician assisting me. I was nervous but very thorough ... very methodical.

"As I dissected the animals that morning, I was blown away, as it were.

"The results absolutely stunned me. I found livers of the first two groups of mice were spotted with gray, cancerous tumors, but the big surprise was the whey-fed group's livers. They remained completely unspotted ... they were pink ... and obviously normal. The colon tumors were significantly less too.

"I was so incredulous and exhilarated at the evidence before me. I just stood there and stared at the awesome results. Very soon, I remembered that I had been trying to help an old Italian lady, a surrogate for my mom, who was suffering with colon cancer. And the next thing I thought was 'Maybe this substance can someday help others like her' ..."

But then he modestly added, "Anyway, perhaps I was only dreaming."

Dr. Bounous remembers that instant as a seminal, almost holy moment of discovery – indeed, perhaps the loftiest in all his professional career. Ironically, once again he felt tremendously alone. Except for his technician, there was no one else to share the experience, and certainly no one else who would entertain the scientific implications as well as he did.

But there was more than science at work. His memory was awakened to touch his heart. Where was Ada Bounous, his angel mother? Could she be in the gallery, clapping and smiling? Was she a believer just like him? Old 'Tito' relished in her approval.

That moment must have felt like a rocket's sharp ascent ... but then came the inevitable drop. After the initial thrill, hard common-sense experience told Bounous that he must calm down. He might tell Luciana, who of course would be glad for him, but his sister could not be expected to understand why this event meant so much to him. Meanwhile, he reasoned, the medical community itself certainly seemed most unlikely to fire off any salutes.

With such ambivalent feelings about what he had just seen with his own eyes, a moment of nostalgia overcame him. Was it too little, too late?

Dr. Kongshavn and probably a few of his other colleagues might well be excited. The rest of the medical world, he well knew, would not be impressed. After all, the Bounous elemental diet had worked very well, providing efficient nutrient support for gut-impaired patients without imposing unwanted side-effects or complications, and at a fraction of the cost of standard intravenous feedings. Still, it had not been widely acknowledged or utilized.

'The elemental diet has proven to be completely effective, yet medicine still resists non-pharmaceutical measures,' he thought. *'No matter what I do with this whey protein, no matter*

how electrifying the results, they won't believe ...'

The cold fact was, there wouldn't be time. At least, not for him. Who knew how many years it might be before the medical world would be ready to look toward common natural remedies for uncommon solutions? Would he live to see his whey protein acknowledged and generally accepted for its major clinical importance? Or would he toil for the rest of his career, possibly achieving even more brilliant results with laboratory mice, yet never seeing his work result in any kind of human healing? The latter application had suddenly become his goal ... **whey protein must somehow have therapeutic use.**

Abruptly, Dr. Gustavo Bounous chose to shut down his thoughts. He would go home, rest and watch some television. He would reward his day's triumphs with a restorative dish of spaghetti and perhaps a light-hearted cartoon.

In the end, the weary scientist had nobody except his girlfriend to tell about his latest shining moment. But more disappointing was the thought that other researchers, in all likelihood, would never understand.

There was so much that Dr. Bounous himself still did not quite understand.

Not yet.

12.

EUREKA!

Dr. Bounous and his co-workers had now made three very important discoveries. The whey protein concentrate or "lactalbumin" which he had been pleasantly surprised to receive initially from Switzerland, had proven to have remarkable effects in the laboratory:

1. The whey protein concentrate had shown a definite, reproducible immuno-enhancing effect in several unrelated strains of laboratory mice.
2. The same whey protein concentrate had inhibited the development of chemically induced tumors in the laboratory mice and somewhat limited their progression or spread.
3. The natural resistance to spontaneous diseases –normal survival –was increased by feeding whey protein to laboratory mice for extended periods just as had been seen for hamsters. In other words, an anti-aging effect was seemingly apparent.

All this drama was totally surprising for Dr. Bounous despite his earlier successes in research on pancreatic proteases. His studies on hemorrhagic shock and the effectiveness of his "elemental diet" had both convinced him of the expanding role that nutrition might play in clinical applications for both prevention and management of disease.

However, after almost a decade of research, this whey protein concentrate had indeed proven to be more than just a nuisance by-product of the cheese manufacturing industry. It was now showing promise for unexpected clinical use; at least for medical research and in the best case scenario, it could have far-reaching implications in clinical practice. What was originally a search for a practical solution to an environmental hazard, was now threatening to become, to some degree at least, an elixir for the human condition. But this would remain nothing more than a prospect, until a more fundamental understanding of the observed effects of the whey protein could be elucidated. That was the puzzle crying out desperately for solution.

The solution was closer at hand than anyone could have guessed.

CHANCE MEETING

A random decision to step down the hall and have a cup of coffee – such an ordinary mid-morning act – turned out to be another one of the most important events of Bounous's career. Was it fortuitous, providential, or one more time, the possible intervention of Tito's mother? Who knows?

The coffee lounge - actually one of the office cubicles which had been outfitted with a few chairs, a small table and a coffee urn - is the last place one might expect to experience an epiphany. With its monotonous brown and tan color scheme, a few shelves of medical volumes and other momentos of the *emeritus* Dr. Fraser Gurd, the ambience is strictly utilitarian – and often quite crowded. God knows, Dr. Bounous needed the social interaction. Laboratory animals must make inadequate companions.

Thus Dr. Bounous seated himself that fateful morning beside a young, blond bearded fellow who turned out to be one Dr. Gerald Batist. The name meant nothing to Bounous, who spent so much time in his laboratory and at his desk that he seldom ventured out to mingle with his estranged but nevertheless illustrious peers. Dr. Batist himself was not much better. He was a private person, a Canadian who devoted much of his extra-curricular time to working with his connections to facilitate similar opportunities for others back in Europe. So although providence had placed their laboratories only yards apart, the meeting of the minds was reserved for this moment in time.

They greeted each other that morning for the first time. As you would expect of avid researchers, they quickly gravitated to shop talk. It turned out that this particular colleague was working on something called **glutathione** and some possible role in cancer.

The mention of glutathione initially meant little to Bounous. That was until he began to describe his own work with whey protein and its immune-enhancing effects on mice, for then Batist seemed surprisingly interested. He sat forward and raised his eyebrows.

"I think that has something to do with glutathione," he informed Bounous.

"Please explain," Dr. Bounous replied instantly, alert and surprised.

The brilliant, young doctor, already established in his field, proceeded to explain how cysteine, a protein building block found in whey, *could* implicate his pet molecule glutathione, the all-important tripeptide synthesized inside the cell for its own protection, and this could lead to immuno-enhancement and more.

Could it be as simple as that? One important molecule? A glutathione connection?

What Batist had just proposed in this chance meeting was as simple as it was profound. Whey must have some unique, or at least special characteristics that could influence cellular glutathione. A lot of research had already been done on that amazing cellular constituent. It was a biochemical gem made inside the cell with a critical role in cell-defense. But no progress had ever been made in its desirably safe, effective or convenient control or manipulation until this time. Perhaps what Batist had suggested could become a key to unlock a treasure of physiological and clinical possibilities. These would hold far-reaching implications of which immunenhancement with increased resistance to infectious pathogens, increased resistance to chemically induced tumors and possible anti-aging effects were only just a beginning.

Bounous sat captivated and amazed as his companion, a quiet man whom he could not recall having seen or talked to before, easily taught him the first possible explanation of why the Bounous/Kongshavn immune system experiments on lab-induced pathologies in mice were resulting in such remarkable success.

"Dr. Batist was *the* expert on *the* subject that I needed to know about. He was the only one at McGill then working on glutathione. There he was, working forty or fifty feet away from me at the door end of the corridor, working on the other half of my experiments without my knowing it," Bounous marveled. "He had the missing link all this time and neither of us knew. We had been so close and yet so far."

"Had this man not known how to measure this," he continued, "my work might have been of no use. He explained to me how he used a spectrophotometric assay. Therefore I felt I had an immediate handle on the problem. At least I could verify his suggestion one way or the other. But my gut instincts told me right there and then that we were on to something."

The two men continued discussing Batist's work during the extended coffee break. He was particularly interested in glutathione in the cancer cell, which he pointed out makes extraordinary amounts of glutathione.

"That morning I told Batist he probably was the most important person in my scientific life!", Bounous exclaimed.

"That got Batist even more interested!"

Elated and reinvigorated by this momentous and most serendipitous discovery, Dr. Bounous proceeded to look again at his whey protein chemistry, and saw that whey indeed contained the cysteine amino acid precursor of glutathione. Some unique property of the whey must make it particularly effective. The enormity of the implications – *why,* ***this whey protein did lead straight to a single main element so vital to cell defense!*** That he realized and it staggered the imagination.

Bounous and Batist were to continue pooling their knowledge very quickly, with the younger man's expertise proving clearly invaluable to Bounous.

"Batist was a warm, brilliant, encouraging man," Bounous said. "Not only was he a true believer in the truth of pure science – an extremely scientific person – but he was also a great philanthropist, a warm and caring individual who succeeded in helping many others immigrate to Canada. This country is so much richer for men like him.

"I owe Gerry Batist a priceless debt of gratitude," Bounous declares.

That unplanned coffee room chat between Doctors Bounous and Batist stands out for Bounous as a monumental leap in his scientific knowledge regarding the protein substance on which he was spending countless research hours. That knowledge later would prove to be not only riveting in importance, but also potentially protective and even life-saving for untold numbers of people.

PROOF

But new knowledge to a critical researcher's mind always stimulates new thoughts and consequently new questions. This leads to the design of new experiments to gain more knowledge which of course, starts the cycle all over again. Again, that's the basic scientific paradigm.

After that memorable coffee break, further discussions and inquiry led Dr. Bounous back to the laboratory. The immediate challenge was naturally to seek to prove Dr. Batist's theory that the immuno-enhancing property of dietary whey protein, in mice at least, was mediated through the ubiquitous glutathione. That he set out to do with great anticipation.

"Gerry Batist was a God-send," Dr. Bounous explained. "He not only had a promising idea, but he had the necessary technique in-house. We could design new experiments to correlate the glutathione levels, which Gerry knew how to measure, with the same kind of immune response assays that I had been doing."

That was an ideal meeting of the minds. They got busy immediately. They designed their study to investigate the mechanism of and the factors responsible for the immuno-enhancing effect of dietary whey protein.

After all, despite the fact that whey protein contains about eight times more cysteine (and only that particular amino acid shows such contrast) compared to casein, there is no elevation in plasma cysteine in the whey protein-fed mice, in comparison to the casein-fed counter part in earlier studies. This was so, despite the observation that the plasma profile of most other amino acids conformed to that of the ingested protein. Where did the cysteine go? Could it not have been used up in glutathione synthesis during lymphocyte proliferation?

The definitive study (33) set out to measure the immune response by plaque-forming cell assay and simultaneously, the splenic glutathione levels. They used three dietary regimes: whey protein, casein and a supplemented casein with added free L-cysteine, in adequate amounts to mimic the cysteine content of the whey protein. They also tried to confirm any possible impact of whey protein on splenic glutathione by measuring the immune response in the presence of buthionine sulfoximine (BSO) which blocks the glutathione synthesis. Furthermore, they investigated the effect of each major component of the whey protein concentrate on the plaque forming cells response.

The spleen glutathione assay was new to Dr. Bounous. They

took carefully weighed samples of mouse spleen, homogenized them in 5-sulfosalicylic acid and then centrituged off the supernatants on the same day. They then could measure quantitatively the concentration of glutathione with an ultraviolet spectrophotometer. This was essentially the method described by Anderson (34) and afforded real tissue values for glutathione in micromoles per gm of wet tissue.

The results were clear. The similar weight gain for each diet group indicated similar food consumption and similar serum protein values were obtained. However, after challenging mice with an immune stimulus and measuring the specific humoral immune response to sheep red blood cells, the response was almost 500% as much for the mice which were fed the whey protein diet compared to the casein diet and the cysteine - enriched casein diet.

The results could not be related to some fortuitous milk protein allergy or some other manifestation of oral immunization. Again, the type of protein in the diet was found to have little or no effective difference on all the other parameters they examined, including body growth, food consumption, serum protein, minerals and trace metals, and circulating white cells (lymphocytes). The difference was therefore not one of "adequacy" of essential amino acids. It had to be in protein type, structure or function.

The only significant effect of protein type was a change in the amino acid profile in the plasma, which conformed to the amino acid composition of the ingested protein, *except for cysteine.* Again, despite the eight-fold higher difference in cysteine content in the whey protein diet, the plasma level of cysteine was not different in the whey protein diet - fed mice from that of the casein diet-fed counterparts. And not surprising, dietary cysteine is known to be the rate limiting precursor for the synthesis of glutathione. Therefore cysteine became highly suspect as the critical variable in the diets.

In the classic experiment by Bounous and Batist, the results demonstrated a significant difference between the effects of casein and whey protein diets on the splenic glutathione concentration during the oxygen-requiring, antigen – driven clonal expansion of the lymphocytes, and following that expansion, in the development of humoral immunity (33). This could reflect the ability of the lymphocytes of the whey protein-fed mice to offset any potential oxidative damage, thus allowing them to respond more fully to the antigenic stimulus. The efficiency of dietary cysteine in inducing supernormal and immune - effective glutathione levels was apparently superior when delivered in the whey protein rather than as free cysteine. Blockage of the glutathione synthesis (with BSO inhibiting a special synthetase enzyme) produced a four to five fold drop in the immune response, just as the Batist theory had suggested. And the immunoenhancing effect is maintained when the whey protein is used and not when free cysteine is added to the casein. Obviously, the specific amino acid profile of whey protein or a cysteine containing peptide is an important factor in determining the fate of ingested cysteine.

So, the theory which was readily advanced by Dr. Batist over coffee that morning when Dr. Bounous happened to sit beside him, was now supported by experimental observations. As Dr. Batist was also quick to point out, glutathione was known to be central in a wide variety of reactions - among them, for example, detoxifying potentially toxic and/or carcinogenic xenobiotics. Thus, they had only just touched the surface. The impact of whey protein by increasing the splenic glutathione levels should have potential implications far beyond the humoral immune response system alone.

And it did.

This was a watershed realization. Dr. Bounous had found the door to open new vistas of clinical possibilities. It was glutathione, the intracellular tripeptide that played such a central role. But more importantly, he realized that he had found the key to open that door. It was as simple and as common as the special dietary whey protein he was studying. He was just beginning to appreciate the potential "medical dynamite" sitting in his research lap.

"Eureka!"

13.

GLUTATHIONE – A CRASH COURSE

Dr. Bounous and his colleagues had made a surprising breakthrough. So much knowledge was accumulating to underscore the importance of glutathione – this ubiquitous molecule that plays such a leading role on the stage of cell-defense. He now had a controlling interest, he thought.

In a short time, Bounous had consumed as much information as he could about his new found treasure. He gobbled up all he could find to read about this key defensive weapon.

If Dr. Bounous himself needed a crash course, most readers of this book can surely benefit from a brief review of some of the basic facts about glutathione. So for the uninitiated in particular, the following is a capsular introduction to this vital cellular component and its activity in the body.

GLUTATHIONE: WHAT IS IT?

Glutathione (35, 36) is a ubiquitous tripeptide molecule, consisting of three amino acids joined together. These are cysteine, glutamic acid and glycine -–three of the twenty two amino acids which comprise the building blocks of all known proteins. In general, the amino-end of one amino acid combines with the acid-end of another to form a peptide bond with the elimination of water. Chains of amino acids are called proteins. The sequence of amino acids and the arrange-

Figure 1. THE CHALLENGE

GLUTATHIONE (GSH) : A Tripeptide (3 amino acids)

γ - GLU - CYSH - GLY

gamma-glutamic acid — cysteine — glycine

GLUTATHIONE SYNTHESIS (inside the cell)

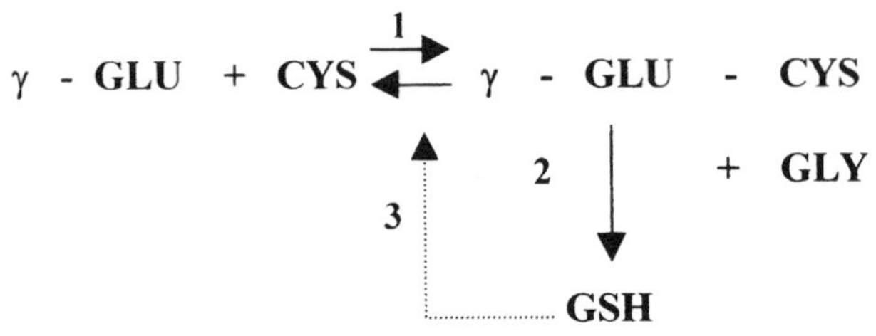

γ - GLU + CYS ⇄(1) γ - GLU - CYS

2 ↓ + GLY

GSH ⋯ **3** ⋯→ (back to reaction 1)

Reaction 1 is catalyzed by the enzyme γ - glutamylcysteine synthetase.
Reaction 2 is catalyzed by the enzyme GSH synthetase.
Reaction 3 is feedback inhibition by GSH, limiting its over-production(60).

The availability of cysteine within the cell is the apparent rate-limiting factor in the synthesis of glutathione to replenish the cell's store during the immune response or after oxidative stress. Therefore, the challenge is to make cysteine available inside the cell for increased synthesis of glutathione on demand.

ment in space of each peptide bond defines some specific structural features of all proteins and oligopeptides (few amino acids in sequence) that relate to their function.

Glutathione (GSH) is only synthesized *inside* cells, in a series of steps catalysed by specific enzymes. The *rate-limiting enzyme* is called γ-glutamyl cysteine synthetase since it joins the glutamic acid and cysteine residues, and the presence of intracellular cysteine is the *rate -limiting substrate.* Fig. 1 shows a simplified reaction scheme.

It is most significant that the amino acids that comprise GSH have so-called thiol (sulfhydryl) groups i.e. – S-H. That is the basis for its abbreviated short form, GSH. The hydrogen atom attached to the sulfur is practically labile and it is this reversible transfer of hydrogen, which itself has an affinity for oxygenated (or oxidized) species, that gives it a powerful anti-oxidant capacity. It typically neutralizes any common oxygenated threat that comes across its path, so to speak. GSH is by far the most prevalent and active intracellular thiol.

Normal cellular metabolism produces a number of reactive oxygen compounds, and inside each cell, numerous mechanisms exist to prevent or treat the possible injurious events that can be triggered by these hyper-active intermediates. Amongst these mechanisms, the "glutathione antioxidant system" is the leading cellular defense. The GSH participates directly in the destruction of reactive oxygen compounds and it also maintains in the reduced (active) forms, other antioxidants such as ascorbate (vitamin C) and tocopherol (vitamin E)(36). Cellular GSH therefore plays a central role in defending mammalian bodies against such insults as infection, free radicals and potential carcinogens (foreign chemicals or xenobiotics)(35). It is also known to function directly or indirectly in many important biological phenomena, including the synthesis of proteins and DNA, transport, enzyme activity and metabolism. Table 2 is a convenient summary of the key functions of GSH.

The multifunctional properties of glutathione become even more obvious by observing the plethora of research publications (21,000+ in the past decade alone) and the wide variety of subjects

TABLE 2. GLUTATHIONE HAS MUTIPLE FUNCTIONS

1. Enhancing The Immune System

Your body's immune activity, involving unimpeded multiplication of lymphocytes and antibody production, requires maintenance of normal levels of glutathione inside the lymphocytes.

2. Antioxidant And Free Radical Scavenger

Glutathione plays a central protective role against the damaging effects of bacteria, viruses, pollutants and free radicals.

3. Regulator Of Other Antioxidants

Without *glutathione*, other important antioxidants such as vitamins C and E cannot do their job adequately to protect your body against disease.

4. A Detoxifying Agent

Another major function of *glutathione* is in the detoxification of foreign chemical compounds such as carcinogens and harmful metabolites.

explored including enzyme mechanisms, biosynthesis of macromolecules, intermediary metabolism, drug metabolism, radiation, carcinogenesis, oxygen toxicity, transport, immune phenomena, endocrinology, environmental toxins, aging and much, much more. Table 3 shows the number of papers which have appeared in a recent decade on glutathione and its relation to a number of clinical conditions. This is clearly a molecular focus for the twenty-first century.

GLUTATHIONE: SUPPLY AND DEMAND

Recall that glutathione is only synthesized inside the cell. Although, in principle, the inflow of cysteine, glutamate and glycine into the cell could prove somewhat limiting under select circumstances, many observations have shown that cysteine availability is indeed the rate limiting factor in GSH synthesis.

Many attempts have been made to enhance glutathione but beside the use of whey protein concentrate, all other approaches have proved futile for a number of different reasons. The major attempts to date would include the following:

(i) **Administration of glutathione (GSH) itself**. Attempts by oral, intravenous, intratracheal or intraperitoneal routes have no sustained effect(37).

Oral (*by mouth*):
GSH is digested or broken down into its amino acid constituents and by themselves, those have no effect.

Intravenous (*by needle*):
GSH has a very short half-life in the circulation.

Intratracheal (*by aerosol inhalation*):
GSH would only affect the respiratory linings, briefly.

Intraperitoneal (*by abdominal wall*):
GSH plasma levels rise but there is no increase in the tissues such as the liver, lung or lymphocytes.

TABLE 3. INTERNET (Med-Line) SEARCH

A Med-Line scan on the Internet shows the widespread research activity now going on to understand further the important role of

- Glutathione in both Health and Disease -

Number of Research Papers in some selected fields, published 1987 – 1997

Aging (957)	Eczema (14)
AIDS (121)	Exercise (171)
Alcoholic Liver Disease (76)	Fertility (37)
Alzheimer's (11)	Gastro-Intestinal (8)
Arthritis (142)	Hepatitis (471)
Atherosclerosis (142)	Infection (406)
Athletics/Aerobics (3)	Lupus (34)
Breast Disease (22)	Malnutrition (79)
Burns (34)	Multiple Scerosis (32)
Cholesterol (391)	Parkinson's Disease (37)
Cancer (3014)	Pregnancy (781)
Cataracts (282)	Prostate Disease (42)
Chronic Fatigue (1)	Radiation (936)
Depression (334)	Stress (2289)
Diabetes (459)	Viral Illness (473)

All this vast body of research points in one direction. Glutathione interrupts common mechanisms that destroy cells and cause many types of degenerative change.

(ii) **Chemically altered glutathione**. Some increases in the levels of glutathione have been noted in specific tissues(38), but the widespread application is limited, particularly because of the harmful or even toxic products of metabolism, such as alcohol and acetaldehyde(39).

(iii) **Amino Acid building blocks**. Oral supplementation with sulfur-containing amino acids such as cysteine and methionine, tend to be associated with toxicity(40). (This is especially so in premature infants, in alcoholics and after surgical stress). Cysteine is also readily metabolized(37).

(iv) **Increasing protein intake**. Nutritional efficiency as measured by the protein content of the diet is unrelated per se to the level of glutathione inside the cells. Commercial whey protein concentrates or cysteine enriched casein prove ineffective in raising the lymphocyte GSH in laboratory animals and the effect is clearly not sustained(41). However, decreased levels of glutathione have been found in many patient groups suffering from protein-energy malnutrition secondary to AIDS, cancer, alcoholism, chronic digestive disorder and burns. This is also a major complication for millions of children in developing countries who suffer from malnutrition (kwashiorkor) – a vicious cycle of disease.

(v) **A few pharmaceutical drugs**:

N-ACETYL CYSTEINE(NAC) is commonly used as an antidote for acetaminophen (tylenol) poisoning(42) and has also been looked at for treatment of HIV and AIDS patients(43). Orally, it has 10% bioavailability. By mouth or intravenously, the effect on glutathione levels is only temporary, and is best reserved for acute care in the emergency room or other critical areas. At higher doses, side effects like gastro-intestinal upset and even cases of anaphylaxis, are not uncommon(44). Case reports of death have already been documented.

Another drug called **L-2-OXOTHIAZOLIDINE-4-CARBOXYLATE (O.T.Z.),** a cysteine precursor as well, has also been used by mouth. It does provide some enhancement of glutathione levels. However, it is subject to feedback inhibition, and to nutritional regulation of glutathione synthesis (37). It therefore does not reliably produce a dramatic increase in tissue concentrations of glutathione.

All the above methods offer at best the interesting possibility for short-term intervention. Their long term effectiveness in providing sustained increases in glutathione inside the cells has still not been confirmed and moreover, there is clearly potential toxicity.

Millions of lives have been saved by specific immunization vaccines, but there is **NO UNIVERSAL VACCINE** for the Immune System. **But Glutathione** is made by all cells for their own internal protection. Is this nature's version of the same idea: **a single, effective agent of defense with widespread efficacy against free radicals, oxygenated species and xenobiotics?**

The GSH antioxidant system is tightly regulated within the cell. Synthesis is increased on demand, while overproduction is limited by feedback inhibition. In effect, *each* cell retains control of its own GSH status. Different conditions may coexist, with each one placing a demand for increased GSH. These might include :

- the production of oxygenated radicals inside the cell during immune response and cell proliferation;
- the metabolic consequences of strenuous muscle exercise such as seen in competitive (or at least serious) athletes, post-exercise;
- the detoxification of foreign pollutants arising from in gestion, metabolism, inhalation or absorption through the skin; and/or
- the deleterious effects of radiation.

It is therefore conceivable that, during such severe challenges at least, there could be competition for inadequate GSH leading to functional deficiencies. In any case, the results Dr. Bounous and others observed in laboratory animals at least, clearly demonstrated the predicted coincidence of the increasing tissue glutathione levels and the consistent beneficial effects of feeding with the unique whey protein concentrate.

This brings us back to our story.

GLUTAHTIONE -ENTER WHEY PROTEIN ... OR EXIT?

As we saw in the previous chapter, a chance meeting during a coffee break had answered a major question that was haunting the mind of Dr. Bounous. *How* did his whey protein concentrate exert such dramatic effects in his laboratory mice? Their immune response was enhanced, their resistance to the initiation and progression of chemically-induced tumors was increased and there was an apparent increase in their natural survival. All this as a direct consequence of feeding these model animals a diet including 20% by weight of the special whey protein that he was blessed to have on hand.

Now he understood much more. These effects were probably related to the increased availability of the rate limiting precursor, the amino acid cysteine, which was somehow made uniquely available to lymphocytes and presumably other cell types when, and only when, the specific whey protein composition (and not even a similar amino acid profile) is used in the formula diets of these laboratory animals.

The fact is that not all whey products are created equal. Bounous was soon to discover a singularly important fact about the white powder which was working so well in his laboratory experiments. Bounous's whey was **undenatured** or *active* protein, a fact he at first did not realize or fully comprehend. He was still emerging from the shadows.

With this understanding of the central role of glutathione in cell-defense and his clear observations that his 20% whey protein diet could effectively raise (or perhaps optimize) the intracellular glutathione levels, Dr. Bounous redoubled his research efforts. There were so many more questions arising almost daily. There was the temptation to jump at short-cut applications before a more thorough understanding of the fundamental mechanisms. But a keen scientist all his life, Bounous refused to be side-tracked from his deepest longing, his real motivation. He must find answers to questions of physiology and biochemistry. How is the effect of whey protein mediated? Why this apparent unique influence? What factors control the glutathione synthesis? ...? Questions and more questions kept arising. This is always typical of the true scientific quest.

Having discovered something with the potential, even if remotely so, for protecting against or actually healing any number of diseases and other misery-producting human maladies, where does one begin to experiment?

At times Dr. Bounous would allow his imagination to take over. He would look ahead confidently to continuing breakthroughs in treating even such intractable diseases as AIDS, hepatitis, chronic fatigue, degenerative arthritis and the like. But above all, Dr. Bounous longed to investigate more fully, and hopefully find even some small success, in preventing or healing human cancers. He thought of his dear mother's suffering, and began daring to hope that he might someday help to alleviate in others, the pain he had watched his gallant champion endure.

God willing, he thought, perhaps ...

But Bounous would quickly realize he was getting far ahead of himself. He was quick to return to reality.

If he could only find more answers to the fundamentals. So, that meant back to the laboratory. Back to designing new experiments and diligently making more observations. He was sure that nature would inevitably yield some other secret that remained obscure, if not to him, then certainly to someone just as daring.

APPARENT FAILURE

Then suddenly, inexplicably ... the unthinkable happened. The whey concentrate simply ceased to be biologically effective. It seemed to have lost all its previous bioactivity. Try as he might, Dr. Bounous no longer found it possible to make any of his experiments succeed. He could not reproduce the previous results. Nothing worked as before. That is a scientist's nightmare. Only the nutritional efficiency of the whey proteins had not changed. There was no longer immuno-enhancement. No bacteria resistance. No tumor inhibition. No anti-aging effect. No more significant effects. Period!

The sudden mysterious and total failure of the substance which had promised such stunning success in the on-going fight against human disease devastated Dr. Bounous. Try as he might, he could no longer obtain any positive results whatever from even one of his many experiments. What had gone wrong? Where had he failed? What monstrous thing had happened? Could this be sabotage? Default or design? Was there no way he could recapture his former successes and regain his professional and personal hopes?

Ironically, now that the scientist at last had discovered why the whey protein worked, suddenly it failed. Trail after laboratory trial failed.

Nearing the peak of his life's work, Dr. Bounous for the first time began to feel hopeless. Was it actually true after all, that his long years of diligent scientific labor ultimately might add up to zero?

It was hard to fight that fear. But Gustavo Bounous ... he did.

1988

"… strong in will to strive, to seek,
to find,
and not to yield."

—Tennyson

14.

PERSONAL TRIALS

Suddenly, it seemed, research had reached an impasse. Experi mental results all turned negative.

It never occurred to Dr. Gustavo Bounous that he might simply quit. At age sixty, with a distinguished and even trail-blazing career behind him, he might have accepted the closed door which now confronted him and decided to retire with grace. But that would bring into question all his reported work during the past decade. Truly the hallmark of good science, is always reproducibility. *Any experiment that's trustworthy must be consistently verifiable, in different hands at different times, given the same conditions.* Bounous knew that maxim and he himself would insist on it. Therefore the onus was on him. He could not retreat now.

Instead, Bounous turned into a bulldog, mentally tossing his whey protein problem in all directions, shaking it, aggressively attacking it – determined to prevail. The perseverance which always had marked his laboratory efforts, the sheer persistence, coupled with a questing and brilliant investigative mind, drew him to advance, not retreat. He would find the answer to why his whey protein concentrate no longer worked, even if it took the remainder of his career.

Actually, it was to take two years - arduous, often exasperating years, during which he methodically crossed off one possibility after another. He tried one idea after another. He searched, questioned, discussed, investigated and tried to unlock what soon appeared to be, a permanently bolted scientific door.

To quit, Bounous deeply believed, would ultimately prove to be a disservice not just to himself, but perhaps also (as he dared to believe) to those with cancer ... AIDS ... pneumonia ... hepatitis ... and a host of other human afflictions against which he longed to eventually utilize whey protein somehow in future clinical trials. He knew his previous results were no fluke. They had clearly offered hope.

FRUSTRATION

However, he also knew that time was running out. His troublesome hip problems could not be ignored much longer. He was now living with pain. Sometimes he would have to force himself to stand for long periods on end as he opened twenty or more animals during the day, examined each in minute detail, and diligently recorded the results of his current observations.

Bounous's answer to such physical and intellectual frustrations was to exert more intensive effort towards learning everything he could about the true nature of that now enigmatic bovine milk whey.

"We knew our techniques were good; the failure was not due to that," he said. He reasoned that something about the whey itself must have changed. Repeated inquiries of the Danish company which now provided his whey powder yielded no clue. The Swiss food company which instigated the initial research had long since moved on to other projects including other ecological solutions.

"The Danes insisted they were sending me the exact same, clean, high-quality whey substance they always had," Bounous incredulously related. "Nothing had changed in the method of production, they told me ... nothing!

"In fact, at one point the company very kindly sent one of their scientists to Montreal to consult with me. He patiently explained every step of the process, answering all my questions, until he convinced me that absolutely nothing was different."

Most people might have felt defeated by such an impasse. In

Bounous's case, however, discouraged and down-hearted as he often felt, he still kept his feet on the path. He showed the same stubbornness which caused him to test his limits, just as he had given all his strength to becoming a surgeon, while knowing that Italy could have no job for him at the end of that weary road.

Tito seemed to be anchored by the same resilient fiber that his angel mother Ada had modeled before him. Sooner or later, *he would find the answer*. He just knew that he knew.

"I did more homework," Dr. Bounous said. "I studied the chemistry of whey and learned it contained four distinct protein elements. I pursued them in detail, chasing any clue that came to mind. But I kept coming up empty-handed. I was scurrying about like a detective but I had no real suspects. Those four protein elements have very different characteristics. Two in particular, I learned, are very labile. Cystine, the direct link to the glutathione that Gerry Batist had suggested is delicate indeed ... So what, I wondered."

Bounous's homework also included consultations with the Dairy Council which is located some forty miles outside Montreal in St. Hyancinthe, Quebec.

"I drove myself there, with my limping hips, at least once a week," he recalls with wry humor. "I traveled more than fifty times to the Dairy Council, and consulted with experts there by phone too many times to count. I felt something might eventually ruboff. Perhaps some light would go on."

Meanwhile, there were scientific articles to write.

"They were all honest accounts of our work and the best ideas that we could come up with at that time." Bounous seemed apologetic.

As he continued to document the important breakthroughs concerning nutrition, whey protein and immunology, the facts linking that interdisciplinary area of knowledge became stronger, more probable ... and, in some ways, more tantalizingly elusive.

"I could not abandon all the strong evidence and meaningful

results we gathered over the years," Dr. Bounous said. "I had always worked hard, hoping to see practical outcome. I knew and had proved that some very significant results already had emerged ... yet now the whey product no longer worked. It was a paradox needing resolution."

Faced with such a dilemma, following years of successful research into whey protein and a stack of authoritative published articles on the subject, some might have been tempted to rest on earlier successes and sweep the later, disappointing results under the rug. They might have chosen to do no more and say no more. But not Dr. Gustavo Bounous.

"Above all, a scientist must never waver from the truth," Bounous declared. "One must remain humble, realizing that setbacks come to all of us. They usually have a disguised purpose. We must never become so arrogant that we turn away from the truth."

If Dr. Bounous believed so ardently in the implacable need for scientific truth, he also believed that even when that truth by some fluke seemed obscured, it could be re-discovered. He was determined to find it.

BLESSING IN DISGUISE

Meanwhile, he labored against some tough deadlines. He had five years before his sixty-fifth birthday and his official retirement from medical research at McGill University, yet he maintained a mental list of projects he desired to complete which would require *fifty* years' work.

Soon the recalcitrant hips absolutely refused to further cooperate. One wintry day in a most unfortunate moment Dr. Bounous fell, a scary situation for someone working and living mostly alone.

"After that fall, I knew it was time to go under the knife," he said. "I could postpone the inevitable no longer."

Dr. Bounous's lifelong love-hate relationship with a solitary life style, both at home and at work, rose up to haunt him now. In solitude one can sometimes find one's strength. But there are times when even the strong become dependent and vulnerable. There is no substitute for family or friends in need. He would need only a minimal amount of care while he recuperated from surgery, yet Bounous the loner could find nobody who would stay in his home and help him. He felt very alone, frightened and discouraged, he recalls. Add such personal distress to all he was carrying in his professional life, and it's easy to understand that he had to fight off depression.

"You've spent your life standing on tiptoe, always on the outside looking in," a psychiatrist colleague told him, not without some sympathy for this man who lived his life as a brilliant accomplished loner. Bounous accepted the verdict: At that age, it seemed most unlikely that his lifestyle would undergo any radical changes.

In the end, Gustavo sent for Luciana. His beloved only sister, he knew, would journey from Italy to attend to him. Indeed, Luciana came quite willingly, spending five months in Montreal while her brother made an excellent recovery from two successive hip surgeries. She was the only real connection he could have to his angel mother's care.

During such dark nights of the soul, Bounous believes the basic building blocks of one's very existence become re-examined and well tested. Anxiety about one's work, future, or the conundrum of living life as a solitary soul, sift down into some recognizable and even comfortable framework. As he says,

"Any 'bad' circumstance is made good if it forces you to seek harder to find the truth."

A loyal sister helps. One's thoughts and ideas continue to move and progress. And there's always television for diversion ...

One night, while tuning out after a day's hard work and alone as usual, Dr. Bounous idly watched a Swiss television program in which two men discussed gourmet cooking. *There* was a program to catch a bachelor's interest. And it did. There was hardly anything unusual about this light entertainment and the few tid-bits of culinary art that the connoiseur chefs delight in passing on to homemakers, hobbyists and perhaps hoteliers. That night, thousands of Montreal TV viewers must have had a routine experience that they would quickly forget. They have enjoyed a steady diet of such programs for years. What could be so special about a simple kitchen fare? A novel recipe? A new twist or technique? Tonight was no different except that all of a sudden ...

"One chef commented that cheese no longer tasted as good as it once did, and the other man replied that when the government required pasteurization temperatures of milk to be raised, it lessened the quality of cheeses."

Those were the first remarks that Bounous could remember from that program. Keen observers hear what they are trained to hear and someone said that '*creative ideas are captured by concerned minds.*'

Amazed and suddenly alert to the import of what he was hearing, Bounous the scientist sat up and leaned towards the television set, straining to grasp every word.

There was more. During the conversation between the two participants, Bounous learned that European milk and cheese producers had been required to raise pasteurization temperatures from seventy two degrees centigrade to seventy-eight degrees, to provide added safety to all European milk and cheese products. This was apparently in response to a previous outbreak of salmonella in France. Bounous and his partner Kongshavn, in the early days you remember, had studied the effect of feeding whey to the mice and noting if their resistance to salmonellosis was altered. What coincidence?

Bounous' mind took off ...

THE GOURMET SOLUTION

Bounous knew - he absolutely knew, with full certainty - that he was near his answer. Alone in his room, again with no one to share a momentous discovery, Bounous's analytical brain leapt from fact to fact as he absorbed the full weight of these new implications.

Salmonella was not the real issue.

Long ago, when he first dug in to study the chemistry of proteins, Dr. Bounous had learned one fact that apparently was now proving to be a major key to the puzzle: the delicate cystine residue in whey protein becomes denatured after it receives a certain amount of heat. *Those few additional degrees of temperature had made his whey proteins incapable of enhancing glutathione in the cells of mice and men!* ... The higher temperature had probably disturbed the cystine structure ... somehow! ... The cystine structure – what's that?

Every high school student today will probably know the story of *The Double Helix*. This is the simple model of Nobel prize winners Watson and Crick who first in 1956 beat out Linus Pauling with the proposal for a three-dimensional description of the DNA molecule. That model consists of two complementary helices, each of which represents the structure of a polymeric strand of nucleic acids, and which are held together by a limited number of specific base – pairs which form intra molecular bridges. That all-important three dimensional structure, as simple as it is and verified by x-ray crystallography and much, much more, forms the basis for the genetic code and explains in meaningful comprehensible terms, a mechanism for such important biological processes as protein synthesis, cell replication, information transfer, cloning and so much more. It is the foundation of modern genetics and cell biology.

That is the sublime example of a general property of biologically active macromolecules. They all have specific composition and structure. That structure derives from the composition of linear repeating units, each of specific geometry, which then give rise to a natural conformation or shape of the final three-dimensional structure. Many important processes are determined by the critical tertiary structure of the complete macromolecule and often their associated units. That explains for example, the ability of hemoglobin to carry oxygen in reversible association in the blood-stream. It is the basis of all enzymatic processes which show a specific key-and-lock stereochemistry, whose unique efficiency leads all inquiring minds to awe and wonder.

That night watching television, Dr. Bounous came to the realization, however serendipitously, that in the production of whey protein concentrate, a change in temperature of just a few degrees during pasteurization of the milk and prior to the casein separation for cheese manufacturing, had had a profound impact on the product. This is not enough heat change to cause separation by boiling off some volatile ingredient or breaking up molecules into various fragments. But it is just adequate to cause labile inter- or intra-molecular bridges to fall apart. In addition, some crucial heat-labile whey proteins could be denatured, hence lost in the curd.

Let's illustrate this important phenomena in even simpler terms.

Think of a chain of beads that is all wrapped upon itself such that its contorted shape (conformation) is retained or held in place by some delicate pieces of thread which make critical connections in specific places. The effect of such small changes in temperature would be equivalent to cutting or dissolving that thread. The consequence is, those pieces of the complete conformation of the chain fall apart.

BIOACTIVITY

That is a conceptual model of what happened to the whey protein concentrate. Just a few degrees change in pasteurization temperature was enough to denature the essential protein structures. The labile disulfide S-S bridges which were so critical could dissolve in just that narrow temperature range. Now Dr. Bounous understood. From batch to batch, a change in preparation conditions – just a few degrees, no more – was all it took to reap havoc in the laboratory. The critical proteins fell apart as it were, and lost that essential bioactivity that had excited Bounous when he first observed the amazing effects on his whey protein-fed mice.

Having been triggered by that incidental gourmet TV program, the once frustrated, depressed Bounous could now grasp the

TABLE 4. PROTEIN COMPOSITION OF COW'S AND HUMAN MILK

Protein	Composition (g/L) Cow's Milk	Human Milk	Cystine/ Molecule
Casein	26	3.2	0*
Beta-lactoglobulin	3.2	Neglibible	2
Alpha-lactalbumin	1.2	2. 8	4
Serum albumin	0.4	0. 6	17
Lactoferrin	0.14	2. 0	17
Total cystine (mol/L)	8.19×10^{-4}	13.87×10^{-4}	
Total cystine (mg/g of proteins)	6.	38.7	

*Casein has 0 to 2 cysteine/molecule

essential elements that came together to make his "mysterious" product work. As he did, his spirits lifted and he rushed back to the laboratory. He now understood what most probably had made the results disappear so precipitously. It was again, just like Dr. Batist's original theory, as simple as it was profound.

The product's bioactivity is dependent upon a critical concentration of three bioactive proteins contained in the milk serum and which are all thermolabile (sensitive to heat). They are :

- ✓ LACTOFERRIN,
- ✓ SERUM ALBUMIN and
- ✓ ALPHA LACTALBUMIN

These proteins contain exceptional amounts of **cysteine**, the critical rate-limiting **molecular** precursor of glutathione. More importantly, it is the **form** of the cysteine which is present that is so critical. The cysteine does occur in pairs in different parts of the protein chains. **Cystine** is the name of this so-called dipeptide unit. Note the different spelling of **cystine** (*the dipeptide*) and **cysteine** (*the amino acid*). In fact, cysteine occurs in two different dipeptide units, i.e. cystine (cys-cys) and glutamyl cysteine (glu-cys). These dipeptides are held together by the disulfide (S-S) bridges which can be kept undenatured under stringent conditions and even remain undigested for absorption into cells. Therein lies the Breakthrough.

When undenatured, these proteins contain almost the same number of cystine residues per total amino acid.(45,46) Hence, as Table 4 shows, in serum albumin, there are 17 cystine residues per 66,000 MW molecule, and six glutamylcystine (Glu-Cys) dipeptides: In lactoferrin, there are 17 cystine residues per 77,000 MW molecule, and four Glu-Cys dipeptides: And in alpha-lactalbumin, there are four cystine residues per 14,000 MW molecule. Conversely, casein, the predominant bovine (cow's milk) protein contains only 0-2 cysteine residues per molecule. Another milk protein beta-lactoglobulin has only two cystine residues per 18,400 MW molecule, and IgG1,

FIG. 2

Synthesis of glutathione: the cell's own antioxidant.
"Immunocal™ as a cysteine delivery system"

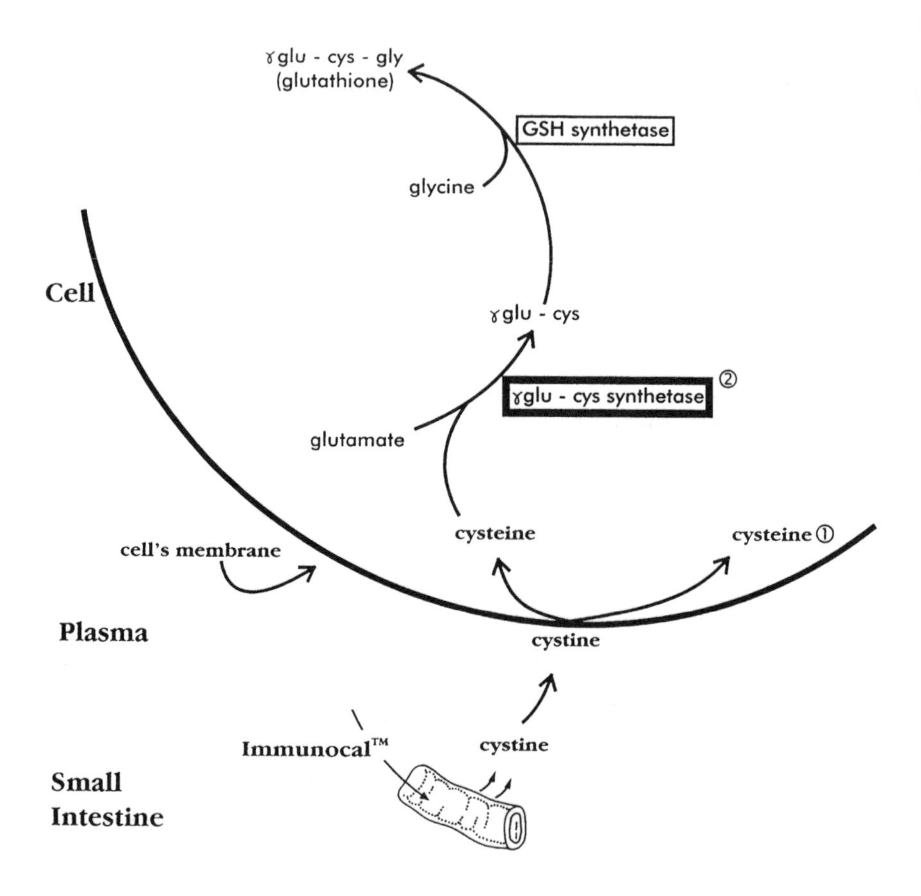

① **Same pathway as the other identical molecule**

② **Inhibited by glutathione levels above normal**

the predominant immunoglobulin in cow's milk serum, only four disulphide bridges (cystine) per 166,000 MW molecule.

In addition, it has been demonstrated that the Glu-Cys precursors of GSH can easily enter the cell to be synthesized into GSH. Interestingly, the Glu-Cys dipeptide is an exclusive feature of the only obligatory foods in the early life of mammals and oviparous species, i.e., milk and egg white respectively (41).

Throughout the digestive – absorptive process, the other co-existing protein fractions of whey (milk serum) influence the rate of release of the glutathione precursors into the blood. This also affords the bio-availability of these crucial ingredients. Uniquely so. Therein again lies the Breakthrough.

Fig. 2 illustrates the simplified process by which Immunocal™ releases cystine or glutamyl cysteine in the small intestine and after transport in the blood plasma, these cysteine dipeptides effectively cross the cell's membrane. Inside the cell, the dipeptide is cleaved to afford the cysteine precursor which can then be utilized for glutathione synthesis.

These facts all came together. There was complete consistency at last. Glutathione was still the door. The whey protein concentrate was still the key. But it could only turn the door knob if the delicate structure of the **cysteine dipeptides** was intact.

- These are the *effective* natural glutathione precursors.

- They (unlike glutathione itself) remain stable in the circulation and efficiently cross the cell membrane barrier to deliver their invaluable cysteine inside.

- They are the essential ingredients that produce significant, rapid glutathione replenishment in lymphocytes during the GSH-depleting immune response in the mice.

- They produce the moderate but sustained increase in organ/ tissue glutathione of old mice following long term administra tion.

- They alone prove to be safe, effective and convenient for that purpose.

- Furthermore, only the *whey protein concentrate proven in Dr. Bounous' discovery could deliver them adequately and consistently.*

- They are **provided in nature** in as common a protein source as the whey derived from milk (and as we shall see, in mother's breast milk), and they are **preserved by science** only with careful, stringent technology.

But our story is getting ahead of itself.

In that fortuitously random chain of events - through mere happenstance – Dr. Bounous found the answer for which he had steadily searched and struggled for two discouraging years. His struggles, he realized, were obviously still far from over, but the worst had been surmounted.

"That television program convinced me that my methods had been right all along ... and proved what I most deeply believe: that one must not take any scientific shortcuts, but must always hold out for the truth," he emphasized. That night he felt vindicated.

FORWARD MOVE

Now Dr. Bounous at last could go forward with his investigations into links between these certain basic proteins, cell nourishment, cell protection, and the resultant potential if all went well, for health protection and the possible healing of both common ailments and perhaps even major diseases.

However, a more formidable problem loomed before him: where on earth could he now obtain again, some good sample supply of whey protein that had not been denatured by the heat of high temperature pasteurization? It would have to be consistently preserved in its native form.

That and other equally challenging parts of the whey protein-to-glutathione puzzle remained. Surely it all must have seemed far beyond one man's abilities to seek out and solve ... even for a man as persistent and resourceful as Dr. Bounous. It was at that point in 1988 that he happened to meet a man who would become an enormously important ally.

Now at age sixty, it would not be very long before Dr. Bounous would have to consider the prospect of retirement from McGill. He would probably have to relocate his research activity if he was to pursue his passionate goals. In the search for a new facility, an approach was made to a prospective landlord who turned out to be far more than a real estate entrepreneur. His name was Dieter Erich Beer. He was not a scientist, to be sure, but an intellectual and rare individual who had excelled in business to an extraordinary degree.

Dr. Bounous at first did not dream that Dieter Beer would turn out to be exactly the right man for the hour, the place, the need, and Bounous's unique scientific circumstances. Another of his life's providential coincidences was about to enter center stage with explosive effect.

15.

THE BOUNOUS – BEER CONNECTION

Another business lunch . . . just one more, among the thousands of working men and women who flood the streets, the parks, the French cafés and small cosmopolitan restaurants in downtown Montreal in the middle of each working day. For many bureaucrats it is a welcome social pastime to interrupt the humdrum and monotony of office towers, computers and paper-work. But for some, it is serious business. It could mean delicate negotiations, deals, contracts, strategic alliances or key decision-making.

That warm summer day, as Dr. Bounous stood in the lobby waiting to greet his would-be landlord, he fidgeted nervously. He was apprehensive about the potential move. Sure he had made some startling discoveries about a whey protein concentrate that now made some scientific sense. It did enhance the immune system – no doubt about that. It did indeed contain valuable precursors to the invaluable glutathione made inside cells, the exploitation of which seemed to open wide doors for application in both health and disease (47). But the discoveries remained far from any commercial application.

So why was Dr. Bounous here to pursue laboratory space in a commercial building? Why even attempt to expand the early work with no funding in place? With limited business experience, how dare he and his colleagues undertake any costly development program? What could anyone do in the market place with a derivation of whey? The formidable challenges of patent applications, human trials, pharmaceutical interests, approvals from Health and Welfare Canada and the US Food and Drug Administration, quality control, etc. all seemed like a vast range of insurmountable barriers that even the experienced would struggle to overcome, much less some naïve medical researchers from academia.

Dr. Bounous remained nervous but excited. He did not have the answers; he did not have the money; but the fascinating discoveries he did have. He could enhance the immune system, demonstrate anti-tumor effect in mice, impact natural aging and generally stimulate these healthy laboratory rodents, all with a surprisingly simple, safe, natural product. There had to be something in this, and it would take only courage to move forward. So here he was in search of office space.

It was not very long before his hosts Ana Damaskin and Dieter Beer stepped out of a city cab soon after it pulled up on the curbside. They were smiling as they entered the lobby and that helped to relax the anxious doctors. But there was more to it than that.

When Dr. Bounous and one of his colleagues sat down to eat with Dieter Beer that day in 1988, he was still hoping to negotiate a lease for some office space in a Montreal building which Beer owned. But Beer's instincts already told him to refuse the request.

"I had many empty offices," he explained, "but I could imagine these doctors would bring sick patients into my office building all day long. It did not seem appropriate."

But Dieter Beer was not the kind of person to be inconsiderate and abrupt. He would not dismiss a prospective tenant or client impulsively. Certainly not distinguished doctors. The opportunity was at least worthy of consideration. Good businessmen are usually open to suggestions and quick to conceive of win-win solutions, or at least compromises. No ideas are bad unless they are final. The world of commerce is dominated by an interconnected network of enterprising people and good ideas. So characteristically, Beer had suggested that they discuss the tenancy matter over lunch.

Sick people walking through the lobby ... Infectious disease carriers on the elevators ... Paramedics in the hallways ... No, regrettably, this would not be good, he decided. His was a commercial office building, not an infirmary.

But as the men conversed during their meal, something else happened. These were not ordinary doctors. They were not running typical patient-care clinics. What they had in mind was something much more fundamental, and much more interesting to the mature entrepreneurial mind that Dieter Beer had cultivated. He was getting a crash course in immunology. Moreover, he was quickly expanding his horizons to appreciate a novel breakthrough in the health and wellness field. He had lots of questions and the doctors were able and quick to oblige him with good answers. Dramatically, Dieter Beer's attitude completely changed. Dr. Bounous's research and its results commanded his full attention. This doctor's ideas and his life's work absolutely fascinated him.

Before the meal ended two and a half hours later, Dr. Bounous had acquired not only the office space he needed, but an immensely important new friend and ally as well. Bounous did not dream that from that day until now, Beer could – and *would* – work miracles in his behalf.

But from that first day, Bounous and Beer instinctively connected.

BUSINESS SUCCESS

Dieter Beer, by then a wealthy Austrian-German businessman, two years earlier had moved with his family to Toronto with the intention of taking early retirement. At that time in his early fifties and financially independent, he thought about returning to school to become a physician. In preference, though, he would accede to the wishes of his wife, Sigrid, and teenage daughters Heike and Britta, who for some time had coaxed him to allow himself more family time. That's what he truly wanted to do.

Beer had dominated the entrepreneurial fast lane since introducing Japanese electronics into Europe in the early sixties. He launched his own electronics import business in 1961 in Wiesbaden, Germany. Acting as the exclusive agent and importer for NEC, AIWA Japan, Daewoo Electronics and Taihan Electronics Korea from 1961 to 1983, he developed and commercialized his own line of audio, hi-fi and video equipment under his company name, TEC. The business he built from ground zero, soon sprawled across Europe and required him to travel extensively throughout that continent, as well as across North America and into Asia. When TEC was sold in 1983, it had been established as Europe's largest importer and distributor in the electronics industry, in addition to selling brand products in North America.

After achieving mega-success in electronics, Beer formed a new business based on his interest in environmental concerns. In 1984 he purchased ORFA AG, a Swiss environmental and engineering company, and subsequently perfected and marketed a fully automated recycling system for mixed solid waste and garbage, which he still jokingly calls the 'Rolls-Royce of recycling.' He implies that its quality and efficiency are second to none, but it's also rather expensive.

"And that's always the trade-off, it seems." Dieter hastens to explain. "You can sometimes cut your corners to hold your prices down. Or else, you hold the standard of quality without compromise and let the consumer choose the value of what you deliver."

Dieter Beer is a true believer in the more excellent way. The licensing of this important technological advance to companies such as Hitachi-Zosen, Kawasaki Heavy Industries, Nippon Steel, ORFA Germany, Multiplan-Benelux and Ecoltecnica Italy etc. was followed by the launching of other ORFA projects in Europe, the Middle East, Japan and Asia.

Obviously Dieter Beer thinks well ahead of others. As he

listened to Dr. Bounous and his colleague over lunch that day, he realized that Europe leads most other parts of the world not only in environmental issues, but also in pharmaceutical and nutritional ideas. Just as Europe provided a perfect proving ground for his *avant garde* recycling business, there were excellent reasons to believe Bounous's milk whey product might also be well received there by those on the forefront of nutrition and medicine. He saw some big potential.

Beer had sold TEC in 1983 in preparation for moving to Toronto two years later. The move was a family decision, he says, and while he retained his Swiss recycling business in Basel, he looked forward to Toronto and the time he'd be able to spend trekking in the Himalayas; or ocean kayaking with friends; or just mountain climbing; or enjoying his hobby of photography, which he took quite seriously; and above all, enjoying his close-knit and very interesting family and their dogs. He had settled down and was just getting comfortable in a new relaxed groove. Then ...

TURNAROUND

... Then came that lunch with Dr. Bounous and news of the immune-enhancing whey by-product that so profoundly intrigued him. Beer found himself asking obvious questions: Who owned the patents? Were there any patents at all? How did they plan to produce this product? How would they market it? For what application or purpose? Who would finance the venture?

"You could turn this into commercial success," he remarked to Dr. Bounous at one point.

"But the fellow I had just met seemed almost distracted at the thought," Beer now remembers. "Obviously he had little idea of what it would take, or how to proceed."

Bounous had spent years developing this potentially marvelous gift to mankind, yet apparently had little notion of what to do with it. Dieter Beer, however knew how to steer a business venture into something as brilliant and dependable as a fine Swiss watch.

"*Why, with some effort* ... ..." His imagination took off ... his mind whirled...his heart raced - just as it had almost three decades before.

That afternoon he told Sigrid his wife about his lunch partners and Dr. Bounous's remarkable new whey protein powder.

"This holds important potential," he told her, as he tried not to observe the look of dismay creeping across her face. He struggled to contain himself. There were promises involved in this. 'Twas a family matter after all.

"I know what we discussed and what I promised," he confessed to his wife," but we must bring this product to market as soon as possible. I *have* to do this."

Sigrid, who knows her husband's idealism and passion very well, could only agree. This was apparently a *cause* for Dieter ... a call to mission. She would be supportive.

Beer's apparently accidental link to Bounous might otherwise appear to be an amazingly fortuitous result of casual table talk ... or else, their strategic alliance might be seen as fully organic as the unseen link between cystine and glutathione – something again providentially orchestrated to be vital and enormously productive, and above all, promising significant human consequences. That irony was too much, but it was real.

Here two men came together from backgrounds that were worlds apart, at exactly the right time and place, to pool their formidable and complementary talents and resources, realizing immediately that they were supposed to move ahead in tandem. They just knew this was the right thing to do. They sensed the resonance and synergy flowing from that first meeting. They determined together to achieve not only continuing scientific proofs and patent protection, but also the means by which Dr. Bounous's pioneering medical-nutritional know-how could eventually be brought to practical use to benefit the typical person anywhere. This must improve the human condition.

Both men could visualize that common objective and even agree on how to approach it, yet neither imagined the incredible number of obstacles they would have to overcome. Perhaps even more risky, neither really knew whether the other actually would persevere until success arrived. Each possessed rare powers of mind, spirit and discipline ... yet, strangely, in the beginning neither knew much about the mettle the other was made of. They connected in trust.

IMMUNOTEC

Immunotec Research Corporation was soon incorporated. The name was conceived by Dieter Beer to recognize the focus on the immune system (**immuno-**), and for historical reasons, to perpetuate his earlier TEC (**-tec**) electronic success. *From its inception,* **Immunotec** *had a noble mission and a grand breakthrough vision.*

The strength of the new Corporation was vested in the originality and promising potential application of the special whey protein concentrate that Bounous had demonstrated was so effective in elevating intracellular glutathione levels. This led to dramatic consequences which already included immuno-enhancement, anti-tumor and anti-aging phenomena at least in mice strains in the laboratory. There could be more to come (47).

Bounous and Beer were quick to recognize the obvious name for the "product" which would now become their focus. It had distinctive impact on the immune system (**immuno-**) which hopefully, would have universal application someday. And it followed in the wake of Bounous' earlier research on 'elemental diets' which had given rise to Flexical™. This name emphasizes that it is a formula diet, and as such, is usually measured in total calorie content (**-cal**). Hence, the name for the new whey protein concentrate became **Immunocal™.**

From the beginning, Immunocal™ had shown such dramatic results by modulating glutathione - the fundamental protective intracellular component, with widespread potential application for both health and disease - that both Bounous and Beer were certain that *they had their hands on a breakthrough product.*

Thus Bounous and Beer embarked on an obviously perilous, always unpredictable and certainly very costly course. And from the outset, they understood that even after working many twelve-hour days and spending a sizeable fortune, they would still have no absolute guarantee of attaining any eventual commercial success. But this project just had to get done, whatever the odds. There was so much at stake.

However, both men *believed.* Bounous believed in the experimental results that he had carefully produced, and Beer believed in the Bounous he had come to know and the potential value of his research. They both believed that what they were about could transcend the mere commercial interests to address genuine human needs. Indeed, they were both totally convinced of the astounding health-giving benefits of the biologically active whey protein product Bounous had developed. They committed to placing their feet as well as their faith on the road to achieving their highly risky and undeniably lofty goal.

'*Only by going out in faith,*' the Austrian-American Dr. Maxwell Maltz philosophized, '*can one see signs and wonders.*' Because they believed, and were prepared to act on that belief, they would later realize 'the substance of things they hoped for and gain the evidence of things they did not yet see.' They were destined to witness the incredible.

VISION

It's action, far more than philosophy, that Dieter Beer's lifestyle portrays. Slender and fit, with European *politesse* and a certain courtly formality in his manner, the Beer style invariably appears calm and unflustered. Only after working with him on a difficult problem, colleagues say, does one begin to appreciate his capacity for steady, unrelenting work.

"I never met a man more determined in the face of adversity," Dr. Bounous declared. "He just would not let me give up. When confronted with a problem, he has the tenacity of a pit-bull. More than once or twice, he saved me from myself. He kept me going at it. He understands only success."

Beer's early education could have predicted his later wide-ranging interests. He studied in Germany, England, France and Spain in the fields of economics, electronics, international law and languages. Beyond that, his global travels and business ventures long ago taught him to think and act beyond conventional borders, perceived boundaries and what others might consider impossibilities.

This is no more exemplified than in his annual trek to the Himalayas in recent years. With small select groups and in the company of only local nomad guides, he pushes himself to the limits of human endurance by hiking into the mountains in the Kingdom of Bhutan in the Himalayas, in a regularly planned, personal pilgrimage. Sixteen to seventeen thousand feet into the rarefied air, with no sophisticated apparatus or equipment – away from all civilization, and uncertain of the next step, much less the next meal – he experiences not just the reserved natural beauty in the silence of the ancient hills, but the authentication of the human experience and the reaffirmation of personal will. And why does he so expose himself to the extremes of personal determination, backed up against the wall of sheer survival?

"You'll never know until you try this," he insists. "It is an experience too beautiful for words. It is an experience of self and of life that defines everything else in relief."

Trekking. This is not a march for dimes. It is a hike for the soul, a march away from dimes and into the depths of human character. It stimulates humility and reflection and engenders reverence and worship. The human spirit climbs to peer beyond the horizon of natural gaze. And one is refreshed, prepared for the next steps of the common human journey.

The experiences are documented not only with vivid mental images for Dieter to recall, but stunning photographs of natural beauty. A prized possession indeed. As we sat comfortably in his ranch office outside Toronto, overlooking the still pond and the verdant rolling hillside country, there was an obvious contrast with the framed photos that cover the rustic walls. They told the stories of Dieter's treks to the precipice of existence and the end of the world. They portray more than the basic necessities of life, they also emphasize its necessary basics. Dieter has mastered them.

CONVICTION

Little did Dr. Bounous suspect that he had joined forces with not only a formidable international business figure, but also a consummate problem solver and a great humanitarian. Further, Beer had acquired an unusual interest in health issues. For that reason especially, he ardently believed he must tackle the difficult challenges and support the cause of bringing Bounous's product to market, somehow. He had deep roots in the wellness movement in Europe which long preceded that in North America.

In fact, more than a century ago, Dieter's grandfather established in Germany the first sanitarium in which herbal and physiotherapy treatments were used to help patients suffering with severe spinal disorders.

The family has had a long tradition of utilizing natural, healthful elements to promote physical and mental well-being. The Beers believed in exercise and good nutrition; maintaining excellent standards of physical fitness; a simple, vigorous lifestyle; and the practice of helping others.

"Dieter Beer was already sold on the importance of nutrition," Bounous said, "and so, he was immediately enthusiastic about the possibilities of the whey product. But more remarkably, his enthusiasm has remained steadfast, no matter how tough things became. And believe me, things did get tough at times."

In a household where Sigrid grinds fresh grains, nuts and dried fruits each morning for the family's breakfast muesli, everyone easily understood Dieter's heightened interest in the new 'wonder nutriceutical.' And even if they felt dismay over the long hours and travels this thing might require, they respected their husband/father's humanitarian reasons alone for pursuing the project. They understood implicitly.

Commend Dieter Beer today for his hard work and very considerable financial investment, and he merely shrugs.

"It's my family who has given," he murmurs.

But his family also received. When they least expected it, an important Beer family concern suddenly came on to center stage. A few years into his work with Dr. Bounous, Dieter Beer's active, healthy mother, then in her late seventies, became ill and had to be hospitalized. She had developed symptoms of throat pain, bronchitis and low grade fever which, on examination, yielded the diagnosis at admission of a right middle lobe pneumonia. Her initial chest x-ray showed an infiltrate behind the cardiac shadow in the right middle lobe. Her doctor showed concern about his. She was worked-up further by CT Scan, and bronchoscopy (including biopsy for histological examination) and lung perfusion scintiphotography (V-Q Scan).

Initially Dieter Beer was told that his mother had "a growth compressing the tip of her lung which caused the pneumonia." In addition, the doctors indicated that the possible "tumor" would not be operable because of her bad heart condition.

Dieter and the rest of the family were devastated by this news and immediately put his mother on Immunocal™. She took it regularly, on a daily basis. Dieter believed in hope that the results Dr. Bounous had seen with this promising whey protein concentrate justified whatever chance there would be of helping his own dear mom. Much like Dr. Bounous, he too was very attached to his mother and would spare nothing to alleviate her condition.

Mrs. Beer was followed up at her doctor's visits. She kept

feeling better and getting stronger.

The fact is that one year later there was nothing to be observed on follow-up examination. There was indeed initial uncertainty as to what caused her condition in the first place. The first x-rays could not exclude a growth of some kind. But after one complete work-up, including biopsy, a malignant growth was excluded and it was determined that the original x-ray finding was most likely a partial 'atelectasis' (lung collapse) of the right middle lobe. Tumor markers were also negative.

Needless to say, Dieter and the entire family were delighted to witness Mrs. Beer's recovery. She rebounded to her active premorbid lifestyle. Whatever the diagnosis, it is most significant that Dieter had experienced a close anecdotal trial of Immunocal™ for the very first time. His conclusions were not scientific by any means, but they dropped an anchor in his heart and soul that reaffirmed his belief in the potential value of Immunocal™ and fortified his commitment to take it to the world.

But there were many bridges yet to cross.

16.

OF MICE, HIV - MEN AND CHILDREN

Immunotec Research Corporation was launched with high and lofty goals. Bounous and Beer were determined to achieve not only continuing scientific proofs and patent protection for the Immunocal™ whey protein concentrate, but they also set out to find some means by which this breakthrough could be applied to benefit the human condition. It was a formidable challenge.

Early studies had shown that the humoral immune response was significantly higher in *mice* fed a diet containing 20 gm of whey protein concentrate per 100 gm of diet than in *mice* fed formula diets of similar nutritional efficiency but where the "bioactive" whey was replaced by other semipurified food proteins(16). It was demonstrated that this immuno-enhancing activity of the whey protein concentrate was related to greater production of splenic glutathione during the oxygen-requiring, antigen-driven clonal expansion of the lymphocyte pool(33). In addition, the *mice* fed the whey protein concentrate exhibited higher levels of tissue glutathione which was believed to account for observed anti-tumor effects(26) and even favorable effects on natural aging(31).

These results were exciting and especially so when the pathophysiology seemed to be related to a common glutathione pathway. The literature was replete with studies on this fascinating intracellular tripeptide. It was known to be an amazing antioxidant, a free radical scavenger, a key co-factor in the effective action of other antioxidants like vitamins C and E, and a major component of detoxifying enzyme systems in the liver(36). Now there was a handle on this ubiquitous cell defender. Indeed one could modulate glutathione with a simple dietary regime that was safe, effective and convenient.

SAFETY FIRST

Yes, but that was all demonstrated in *mice*. Many a promising pharmaceutical product has been aborted in transition from laboratory bench to laboratory animals or from laboratory animals to human subjects. For a wide variety of reasons, a solution that works *in vitro* (test tube) may fail *in vivo* (live animal) or may be totally inapplicable for human application or treatment. Perhaps the human pathophysiology turns out to be different, or the side effects are awful, or the drug stability is inadequate, etc. So there could be big hurdles to climb in the shift from test tube, to laboratory animal to human beings.

But Bounous's whey protein concentrate was different from the outset. Here was a totally natural product wherein the active proteins in common with mother's breast milk were made available in concentrated form. What could be more natural? The effect was essentially to enhance glutathione production in cells by an intrinsic process which was itself regulated within *each* cell by feedback inhibition.

So the question of safety of the whey protein concentrate seemed redundant, although one could never be sure. But still, one should never rush into human trials without due scientific consideration and review of all the available information. And protocol must be deliberately and systematically applied. Dr. Bounous believed in his nutritional innovation but he wrestled with these considerations and more. He would only countenance doing the right thing in the right way and at the right time. He was a true scientist with deference for the clinicians.

THE JAPANESE CONNECTION

Dieter Beer himself was thinking and seeing beyond the immediate horizons. His frame of reference was global. His experiences were rooted in Asia so he conceived of even distant associations. Any kind of clinical trials or pharmaceutical development, which was his initial perspective, would necessitate some major player or players. He understood so well the synergic power in strategic alliance. So he wanted to explore whatever contacts he knew.

"You have to understand that I retired from McGill University to work out of Dieter's office building, only to have him push me to work harder." Dr. Bounous was not complaining. He knew his natural tendency to take the paths of least resistance. He was grateful for such a push. The project needed it. "Dieter believes in travelling to the ends of the earth to accomplish his mission. I hate to travel and warned him that I usually refused to go anywhere."

Two weeks later Dr. Bounous was holding in his hand an airline ticket to Japan. The trip to Osaka came through Beer's contacts with a large Japanese pharmaceutical company which was broadly involved with nutritional products. The Otsuka Pharmaceutical Company Limited had an active Cellular Technology Institute in Osaka which was a major player in the nutritional field.

"They liked Mr. Beer and seemed to like me, and exhibited an interest in doing a little research on my project," Dr. Bounous is so appreciative of every contributor along the way.

"In 1990 they had sent Dr. Khori, their Director of Nutrition, with four junior executives, to our Montreal office. We had no product at the time, but these Japanese gentlemen became aware that we were trying hard to approach production. Though they had made a long trip to evaluate what we had, at that time we could give them absolutely nothing but words and promises. But that's the Japanese style: they buy-in to people and follow their instincts with trust and loyalty.

"They believed in us, and returned to Montreal at least four times. During that period we had a small Quebec dairy producing our whey product on an irregular basis. Later we got great support from a government agricultural station in St. Hyacinthe, Quebec".

In fact, between 1988 and 1990, they could only produce 3-5 kg of Immunocal™ per month and this with the help of the Canadian and Quebec governments' food research institute in St. Hyacinthe. Since meaningful clinical trials with humans were not possible with these quantities, they had to find the possibility of producing up to 100 – 200 kg per month. This led to lengthy research and discussions with St. Hyacinthe scientists, Dr. Bounous and European manufacturers of the most modern equipment. Dieter Beer finally leased the necessary pilot plant facilities for close to two years. They were installed on a temporary basis in the St. Hyacinthe research institute. This then allowed the production of 20-30 kg of Immunocal™ each month and again, with the help of the same European equipment manufacturer, they found a small Canadian dairy, the only one in North America at the time, which had the technical capability for Immunocal™'s pre-production. Upon recommendation of their scientific consultant, they agreed to install equipment and machinery which would allow production of 200-300 kg per month.

From 1993 onwards, Immunotec therefore had the capability of producing on contract, the necessary quantity of Immunocal™ to work on perfecting mass production technology, improving production quality and experimenting with shelf-life. More importantly, there would be enough product for increasing clinical trials in Canada and overseas, especially in Japan. They could supply a large number of individual concerns who were willing to collaborate on a long term study with Immunocal™. The group at St. Hyacinthe persevered until they could deliver adequate quantities of the good quality whey product that produced the dramatic results they had come to expect earlier.

"I had no significant laboratory then", Dr. Bounous pointed out, "but still those cordial, loyal Japanese people continued to conduct research with our whey concentrate. It was like having your own lab ten thousand miles away. They would conduct research, then come and bring me the results."

"Dieter obliged me to go to Japan to meet Mr. Otsuka, an old, kind man whose father had founded the pharmaceutical company. We visited this elderly son who had built his father's business into the second largest such firm in Japan. We met him at his little home by the sea. It was an oriental treat. The experience of meeting him in such relaxed ambience could almost justify the ordeal of the trip."

In the early days, Bounous says, they could not even give the Japanese a guaranteed consistent product. Not only did they lack sufficient quantities of whey concentrate to provide for large clinical trials, but the product itself, or at least the methods of producing it, continued to change. Somehow, the Japanese were committed enough to hang in with them. Eventually they got past the hurdles and later the facility in St. Hyacinthe was able to meet the demands.

"Deiter and I were like two pygmies working with that pharmaceutical giant," Dr. Bounous remarked. "They remained intelligent, professional and extremely loyal. I was impressed at the lengths they would go to help bring our product to fruition."

The first results coming out of Otsuka were very important and heart warming for Dr. Bounous. The group there showed that spleen cells of a select strain of male mice fed a 25 gm Immunocal/ 100 gm diet for 4 weeks had an increased immune response to sheep red blood cells (sRBC) *in vitro*, and a higher content of L3T4+ cells than mice fed on an isocaloric diet with 25 gm pure casein/100 gm diet. Similarly, the spleen L3T4+/LYt-2+ ratio was two and a half times higher in Immunocal-fed mice compared to the casein-fed controls. That was a definitive demonstration of the immune enhancement by Immunocal™, in different hands at a different time and in a different place. Bounous was reassured.

That report(48) published in December 1990, was actually the first unequivocal, independent duplication of Bounous's earlier experiments. It preceded the subsequent results of the CSIRO Laboratories in Australia. At the time it was confirmatory proof-positive of the earlier discovery and it boosted the confidence of the Montreal group to push ahead and consider even a pilot case human study.

HUMAN TRIALS BEGIN

The choice selection for such a study came from yet another providential circumstance. Enter Dr. Sylvain Baruchel ... a brilliant, young pediatric oncologist connected at that time to Montreal Children's Hospital situated but minutes away from Dr. Bounous's stomping ground. He read Dr. Bounous's work in the literature and was fascinated by it. He had a special interest in clinical immunology, having worked with Dr. Luc Montagnier at the Pasteur Institute in Paris. This is the Dr. Montagnier who just a few years earlier had been credited with the discovery of the virus responsible for the AIDS epidemic.

Creative ideas are known to gravitate to prepared minds. Dr. Baruchel was prepared. He spotted early the implication of the Bounous discovery, at least for his own field of interest. So the keen, energetic Dr. Baruchel came to Dr. Bounous in 1990, immediately proposing that his obviously nutritious and perfectly harmless whey powder be used in a small pilot study, *to investigate the possible beneficial effect of Immunocal™ in sympton-free HIV-seropositive individuals.* After all, there is no better proof of efficacy in the immune system and no greater need in recent time than the possible effect on patients with human immunodeficiency virus (HIV) or acquired immune deficiency syndrome (AIDS). The syndrome is characterized by gross immunodeficiency, low T-helper cell blood content, increased oxidative stress and ... yes, systemic glutathione deficiency.

Dr. Baruchel was one of the first to outline the role of oxidative stress in HIV/AIDS disease progression and the potential for the use of antioxidant in HIV disease(49, 50). Oxidative stress occurs when the balance between free radical generation and antioxidant defense is upset. It is a known factor of HIV replication *in vitro* and has a potential role as a co-factor of HIV disease progression.

In human terms, after primary infection by the virus and viral dissemination, most patients have a period of 'clinical latency' that may last for years(51). Then comes the decline of active immunity and the downhill pattern of the wasting syndrome experienced by the vast majority of AIDS patients. The immune system cries, then it dies.

The varied factors that stimulate HIV to replicate and to determine the period of latency are still poorly understood *in vivo*. But Buhl and co-workers had recently documented the ominous systemic glutathione deficiency in sympton-free HIV-seropositive individuals(52). Dr. Baruchel reasoned that there was therefore only positive indication for a small Immunocal™ pilot study. It was time to begin. A protocol was soon devised.

AN AIDS PILOT STUDY

Immunocal™ was given orally to 3 male HIV-seropositive individuals, ages 29-35. These patients took the product daily in a liquid of their choice for a period of 3 months. The daily intake of pure whey protein prescribed to the patients as Immunocal™ was increased step-wise. During the first 4 weeks, 8.4 gm were prescribed daily; in the following 4 weeks, 19.6 gm; and in the final 4-week period the dose was raised to 28 gm (first week) and 39.2 gm (last 3 weeks). Protein intake from other sources was reduced by corresponding amounts(53).

Dr. Baruchel made a major contribution at that time in developing a reliable *in vitro* assay for routine measurement of cellular glutathione levels. This was critical for patient response monitoring.

Well, what of the results of this initial pilot study?

Three patients took Immunocal™ daily for the 3-month period without any adverse side effects. In all these patients, body weight increased progressively (from 2 to 7 kg); 2 of them reached ideal body weight. It is interesting to note that the body weights of all three patients were stable for at least two months prior to the study. Serum proteins, including albumin, remained unchanged and within normal range, indicating that protein replenishment per se was not likely the cause of the increased body weight.

The glutathione content of blood mononuclear cells was as expected, below normal values in all patients at the onset of the study. Over the 3-month period, however, glutathione levels increased and in one case, it rose by 70% to reach normal value.

These objective changes were accompanied by a marked improvement of a subjective sense of well-being in all three patients.

It is noteworthy that one patient, unduly concerned that the beneficial increase in body weight could hamper his preferred lean appearance, drastically reduced his Immunocal™ and total energy intake during the second period of study. During this time, body weight increase was reduced and glutathione failed to rise. Three comparable patients on their usual standard diets over the same period, showed some weight loss and no change in their blood GSH mononuclear cell content.

So what could one conclude from this?

The preliminary data indicated that whenever patients maintained their energy intake at pre-study levels but replaced a significant portion of the protein intake with Immunocal™, body weight increased and mononuclear cell glutathione increased. Given that cellular glutathione is very tightly regulated and that the pre-study cellular glutathione values were very similar in the three patients, the observed increases in cellular glutathione concentration were likely to have biological importance.

The positive effects of Immunocal™ observed in that very limited number of HIV-seropositive individuals acquired significance when viewed on the background of the large number of animal experiments showing elevation of cell glutathione and immune response on Immunocal™. Animal studies emphasized the fact that the immuno-enhancing effect of Immunocal™ was not related to a greater systemic nutritional efficiency when compared to several other protein sources with similar nutritional efficiency but no significant biological activity. Mice fed Immunocal™ did not exhibit increased body growth or any changes in serum protein levels. Similarly, in these patients Immunocal™ did not produce any change in serum proteins, which remained constant throughout the study. The increase in body weight observed in the patients did not correlate with increase in energy or protein intake throughout the study period but rather with improved sense of well-being. The extra protein intake through Immunocal™ was generally compensated by reduced intake of protein from other sources. The whey product was well tolerated in all three patients, at different doses, again with no side effects.

Earlier laboratory studies indicated that whey protein concentrates, from other sources, did not produce significant biological activities while exhibiting similar nutritional efficiency. Immunocal™ was indeed a unique product.

This preliminary study(53) published in 1993 clearly indicated the need for further clinical investigation on the effect of Immunocal™ in HIV-seropositive asymptomatic or symptomatic patients. The whey proteins, by providing specific substrate containing cysteine for glutathione replenishment in the lymphocytes, could indeed represent an adjuvant at least to other forms of therapy(54).

These were remarkable results considering the alternatives available for HIV-seropositive individuals. This was even before widespread triple-drug therapy.

CHILDREN WITH AIDS

Dr. Baruchel became very interested in initiating a similar study in HIV-seropositive children. He led the Montreal group to conduct a Canadian clinical trial (Canadian HIV Trials Network) with Immunocal™ in children with AIDS and wasting syndrome. The major objective was to evaluate the effect of oral supplementation with Immunocal™ on nutritional parameters and intracellular blood lymphocyte GSH concentration in such children. This was an open single-arm pilot study of 6 months duration. Wasting syndrome and severe weight loss within the 6 months preceding entry into the study was an absolute criterion for entry.

In this protocol, Immunocal™ was administered twice a day as a powder diluted in water. In some patients, Immunocal™ was administered via nasogastric tube when necessary. The administered starting dose was based on 20% of the total daily protein requirement and was increased by 5% each month over 4 months to reach 35% of the total protein intake at the end of the study. The total duration of the study was 6 months.

The children were monitored regularly for their clinical respone to the special diet. Weight, height, triceps skinfold and mid-arm muscle circumferences, CD4/CD8 counts, and peripheral lymphocyte GSH concentrations (measured by spetrophotometric assay) were measured monthly. Energy intake was assessed by the use of two independent 2-day food records with a 2-3 week period between the food records. Each food record included a weekday and a weekend, and the average of these records was calculated to reflect the daily nutritional intake. Out of 14 patients enrolled, 10 were evaluable. The ages of the patients ranged from 8 months to 15 years. The 10 patients studied were enrolled in four different centers across Canada: Montreal Children's Hospital (Dr. S. Baruchel), The Hospital For Sick Children Toronto (Dr. S. King), Children's Hospital For Eastern Ontario (Dr. U. Allen), and Centre Hospitalier Laval Quebec (Dr. F. Boucher). Of the 4 remaining patients, 2 lacked compliance after 2 months while the other 2 died of AIDS progressive disease within the first 2 months of entry into the study. None of the deaths was related to the tested product.

None of the patients experienced any major toxicity such as diarrhea or vomiting or manifestation of milk intolerance. One patient chose to stop Immunocal™ transiently for minor digestive intolerance such as nausea and vomiting (< twice/day) at month 3 and was subsequently able to restart the treatment without any problem.

At the end of the study, all patients experienced a weight gain in the range of 3.2% to 22% from their starting weight. The mean weight gain for the group was 8.4% ± 5.7%. Recall that this was a reversal to the severe weight loss trend immediately prior to the study. On analysis of the mean percentage of requirement nutrient intake (RNI) per month for all the patients, no correlation was found between the weight gain and any significant increase in the mean percentage of RNI, suggesting reduced catabolism rather than an anabolic effect of Immunocal™. Six of ten patients demonstrated an improvement in their anthropometric parameters such as triceps, skinfold or mid-arm muscle circumference independently of an increase in energy intake.

Two groups of patients were identified in terms of GSH (glutathione) modulation: responders and nonresponders. The responders were those who started the study with a low GSH level. The nonresponders were those who started with a normal GSH level. A positive correlation was found between increase in weight and increase in GSH. No changes were found in terms of blood lymphocyte CD4 cell count, but two patients exhibited an increase in the percentage of their CD8 cells and four patients showed a trend toward an increase in the number of NK cells.

In conclusion, this pilot study(55) demonstrated that Immunocal™ was very well tolerated in children with AIDS and wasting syndrome and is associated with an amelioration of the nutritional status of the patient as reflected by weight and anthropometric parameters. Moreover, the GSH-promoting activity of Immunocal™ *in vivo* seemed to be validated in six out of ten patients.

A multicenter double-blind randomized study is currently under way in Canada in adult patients with AIDS and wasting syndrome.

These results are consistent with a more recent discovery by Herzenberg and co-workers of the crucial importance of maintaining the GSH content of CD4 helper T-cells for the survival of HIV-infected patients(56). Hence, as was shown by Herzenberg in the area of survival, Baruchel's data substantiate the crucial correlation between GSH repletion and the patient's clinical improvement, i.e. cessation of wasting and/or increased body weight, unrelated to calorie intake.

In a joint research project with Dr. Mark Wainberg from the McGill AIDS Research Center, Baruchel demonstrated further that Immunocal™, functioning as a cysteine delivery system, enhances GSH synthesis by mononuclear cells and inhibits HIV replication *in vitro*, as measured by reverse transcriptase activity in an infected cord mononucleated cell system. As well, Immunocal™ was found to inhibit the formation of syncitium between infected and non-infected cells and reduce apoptosis (cell death) of HIV infected cells. The inhibition of syncitium formation occurred at the same concentration as the inhibition of HIV replication. These results(55) were crucial to the understanding of the mechanism underlying the effect of Immunocal™ feeding in AIDS patients. The favorable clinical response could then be attributed to a direct anti-viral effect of raised GSH levels induced by some of the specific whey proteins.

In a further joint research project with Dr. René Olivier at the Pasteur Institute in Paris, Baruchel demonstrated that HIV infected cells from AIDS patients when fed *in-vitro* with Immunocal™ would not die prematurely and would survive longer(55). These results generated major interest and were underlined by Dr. Luc Montagnier in the opening ceremony of the Tenth International AIDS Conference in Yokohama, Japan "as an important new line of research which should be expanded."

The *in vitro* assay of GSH synthesis by mononucleated cells developed by Baruchel was also used to verify the bioactivity of different whey protein concentrates. Immunocal™ outperformed all others as seen in Fig. 3. The same assay was also used for quality control of different future batches of Immunocal™. With these labile proteins, it is most important to safeguard consistent quality for reproducible results.

The results substantiated the proposal that normal GSH

FIG. 3

S. Baruchel utilized the *in vitro* assay to compare the glutathione promoting activity of most currently available serum milk protein products(55).

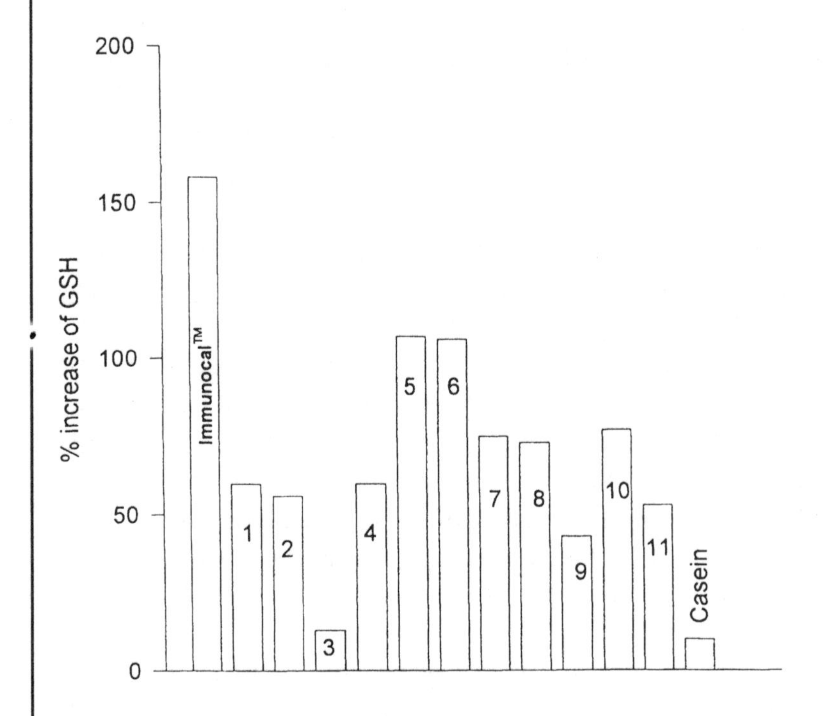

Incubation of PBMCs for 72h in the presence of Immunocal® and other serum milk products: Percentage increase in glutathione.

values in the mononucleated cells prevent HIV replication(57). On the other hand HIV infection per se may directly affect GSH, as it decreases the activity of glucose – 6 phosphate dehydrogenase (G6PD), a key enzyme in the pathways that maintain GSH in its reduced state(58). Pulse administration of the cysteine derivative N-acetylcysteine (NAC) increased GSH blood level and survival in one study and plasma cystine in another study(43). CD4+T-cell numbers did not, on the average, change significantly after N-acetylcysteine treatment(43).

Other case studies of HIV-infected individuals continue to accumulate. The best example of the value of Immunocal™ to such patients is illustrated by a family of three. They are included in a short series of case reports which now follow.

AIDS - CASE REPORTS

The following case reports illustrate the effect of providing patients with Immunocal™ as a natural cysteine delivery system as the only treatment of HIV infection over a 6-month period. Most of these patients had not been able to tolerate the adverse effects of antiretroviral therapy. Only patients that had documented laboratory data were selected.

Firstly, the family of three.

1. The Father

This previously healthy 46 year old man was hospitalized in March 1995 for bacterial pneumonia that resolved with antibiotics. In April 1995 he was diagnosed HIV-1 positive and developed bilateral axillary adenopathy. He was infected through a heterosexual extra-marital relation. In May 1995, he was treated with AZT for one week but the drug was discontinued owing to strong adverse reactions. He was constantly fatigued and had to abandon his job. No wasting was seen.

In April 1997, oral treatment with Immunocal™was initiated, 25 gm a day. Two weeks later the patient felt stronger and was able to resume his work on a regular basis. The lymphoadenopathy had significantly improved. However, in early August 1997, he was advised to take 5 gm a day of vitamin C. A few days later he experienced fatigue and for the first time in 4 months he was forced to quit his work. He discontinued vitamin C intake on September 10th . Two weeks later his general condition had improved and he was feeling stronger. The time course of laboratory investigation is shown in the next table:

Immunocal™ – From April 1997

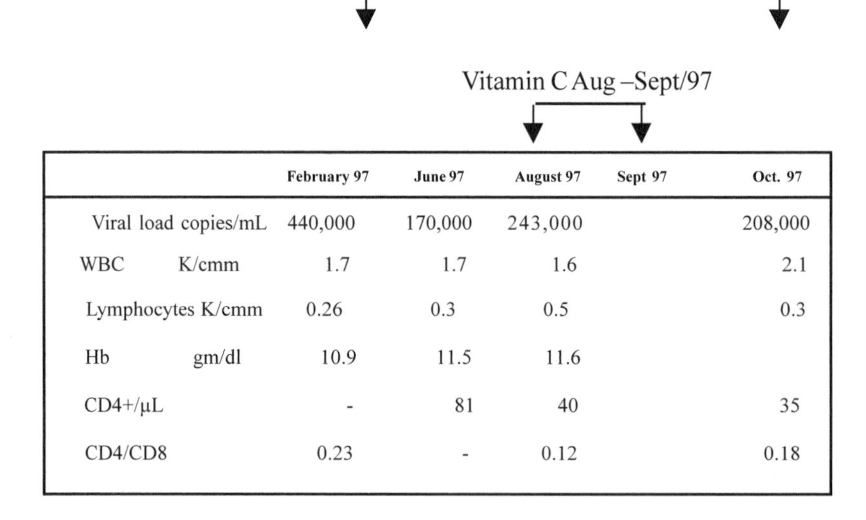

	February 97	June 97	August 97	Sept 97	Oct. 97
Viral load copies/mL	440,000	170,000	243,000		208,000
WBC K/cmm	1.7	1.7	1.6		2.1
Lymphocytes K/cmm	0.26	0.3	0.5		0.3
Hb gm/dl	10.9	11.5	11.6		
CD4+/μL	-	81	40		35
CD4/CD8	0.23	-	0.12		0.18

2. The Mother

This 35 year old woman was diagnosed HIV-1 positive a few days after her husband was tested positive in April 1995. She presented at the time with right cervical lymphoadeopathy. She tried AZT for one month but was forced to stop it because of vomiting and an intolerale headache driving her to the point of suicidal ideation.

This incidentally was the reason why she refused AZT treatment for her son. No weight loss was documented.

She was started on Immunocal™ 20gm a day, in April 1997 and 2 weeks later noticed increased energy and strength. Her lymphoadenopathy cleared and has not reappeared since. Her time course of laboratory investigation is again shown in the following table:

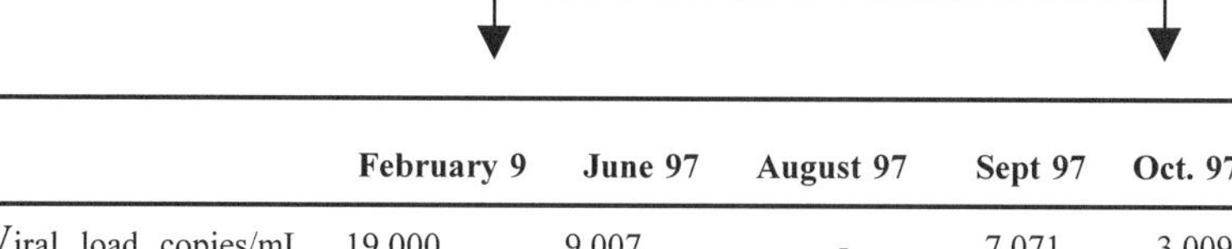

	February 9	June 97	August 97	Sept 97	Oct. 97
Viral load copies/mL	19,000	9,007	-	7,071	3,009
WBC K/cmm	3.0	3.1	5.1		6.8
Hb gm/dl	11.4	11.8	12.4		
CD4+/µL	410	420	400		
CD4/CD8	1.32	1.31	1.35		

This young woman showed a progressive decline in the virus load and an increase in WBC with slight Hb increment. These positive lab data are associated with a major improvement in strength and sense of well being, persisting after 6 months of therapy.

3. The Son

The 2 year old boy was tested positive about the same time as the parents in April 1995. He did not appear to suffer major inconvenience and he exhibited no failure to thrive. He was started on Immunocal™ (10gm daily) from April/97. He showed increased energy in playing and six months later he was doing very well. Because of the magnitude of the improvement after only 2 month therapy, the October cell blood count results were verified at the "Fletcher Allen Health Care" in Burlington, VT and the Mayo Clinic Laboratory. The time course of laboratory investigation is in the following table.

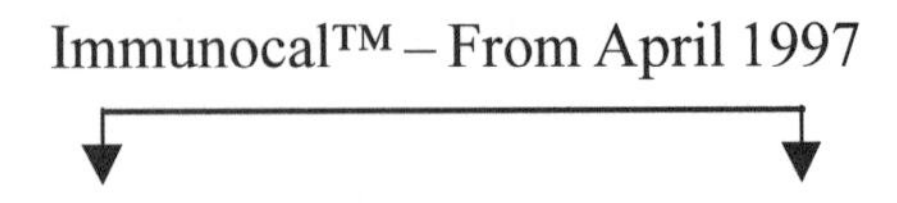

	February 9	June 97	Sept 97	Oct 97
Viral load copies/mL	140,000	1,104	5,000	
WBC K/cmm	4.7	4.8		7.23 cf. Sept 95 value: 7.3 (at age 15 months)
Lymphocytes K/cmm	2.5	3.0		3.33
Neutrophils K/cmm	1.6	1.1		3.65
CD4+/µL	1,025	1,200		1,450
CD4/CD8	1.78	2.35		2.2

Up to the time of writing, this entire family continues to show steady clinical progress while continuing on Immunocal™. The mother in particular has since shown a very positive response in a crossover trial of triple drug therapy. It appears that Immunocal™ replenishes the glutathione consumed by the standard drugs, thereby improving patient tolerance. Further studies are indicated.

4. This 44 year old homosexual male was first tested HIV-1 positive in 1986. In the following years he felt progressively weaker, had recurrent episodes of diarrhea and frequent skin lesions such as eczema and herpes. He always refused antiretroviral drug therapy. No wasting developed. He began Immunocal™ treatment on February 1, 1997 at a daily dose of 10 gm. Two weeks after, he noticed increased energy and physical strength which has remained. He has not experienced diarrhea during the last nine months and the skin lesions have greatly improved.

The time course of laboratory investigation is shown in the next table:

Immunocal-Feb 1997 ff.

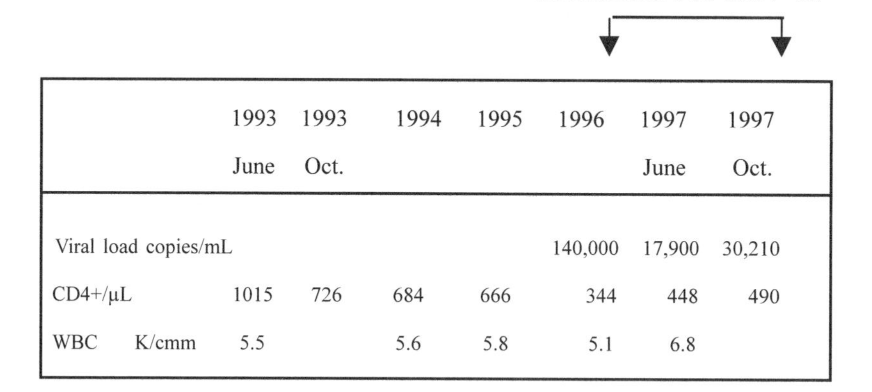

	1993 June	1993 Oct.	1994	1995	1996	1997 June	1997 Oct.
Viral load copies/mL					140,000	17,900	30,210
CD4+/μL	1015	726	684	666	344	448	490
WBC K/cmm	5.5		5.6	5.8	5.1	6.8	

The few cases reported here show that the administration of Immunocal™ can improve the clinical condition of HIV-infected patients. This favorable response may be associated with a decline of the plasma virus load reflecting lowered virus production in the lymphoid tissue. In the mother, CD4 levels remained steady throughout. The best results are seen in the 2 year old boy. The results in such a young child and the Baruchel study in children obviate the likelihood of a placebo effect. However, at this time, ***this form of dietary intervention should be viewed as complementary rather than an alternative to antiretroviral drug therapy.***

GSH maintains vitamin C in its reduced functional state. Hence depletion of GSH could occur in AIDS patients during megadose administration of vitamin C. This would explain the clinical deterioration observed when megadoses of vitamin C were administered in patient #1. It is noteworthy that Herzenberg et al. cautioned AIDS patients to avoid excessive amounts of products likely to alter GSH levels such as vitamin C and vitamin E(56).

CONLUSIONS

Undenatured whey proteins and more specifically, the cystine-rich thermolabile proteins, represent *an effective cysteine delivery system* for the cellular synthesis of glutathione. The oral administration of these proteins (Immunocal™) induces a rapid replenishment of glutathione (GSH) in the lymphocyte during the GSH-depleting development of the immune response. Less hampered by oxyradicals, the lymphocyte can thus fully react with an optimal immune response to the antigenic stimulus.

Similarly, the low GSH level in the lymphocytes of AIDS patients can be reconstituted to normal values by *oral administration* of Immunocal™.

The demonstration by Watanabe et al.(59) that Immunocal™ increases GSH levels and inhibits the hepatitis B virus, supports the following hypothesis: the long incubation of the HIV virus until GSH levels drop in the lymphocyte hosting it, suggests that *the HIV virus cannot adapt by mutation to a normal GSH level as it would to external drugs*, because GSH is a natural component equally antagonistic to other viruses such as the viruses that cause hepatitis.

Immunocal™ might be viewed as a "natural" product with regard to its origin because it is derived from bovine milk and does not contains additives. However, its preparation involves the most advanced technology of micro-filtration, ultra-filtration and drying aimed at preserving the labile milk serum proteins in their natural form. That affords its *essential bioactivity*, acting as a cysteine delivery system for the cellular synthesis of glutathione.

Cellular glutathione (GSH) is a tightly self-regulated system because of the *feedback inhibition* of gamma-glutamylcysteine synthetase activity by the GSH level[60]. However, when GSH is depleted, as in the lymphocytes of mice during the immune response or in the lymphocytes of AIDS patients, the cysteine delivery system in Immunocal™ produces a substantial increase in cellular GSH up to but not above normal values. Preliminary data in AIDS patients demonstrate that this is associated with major improvements in health.

These clinical data and the *in vitro* demonstration that Immunocal™ inhibits the HIV virus while increasing GSH synthesis, strongly suggest that *an antagonistic relation exists between the viruses and cellular GSH.*

Unlike specific antiretroviral drugs which may induce mutation (hence resistance of the virus to therapy), the normalization of the lymphocyte glutathione levels and redox status through a cysteine delivery system, represents *a totally different approach* by which the natural cellular defense system is boosted and against which the virus cannot apparently build up resistance by mutation[57].

But the relevance to viruses and HIV/AIDS in particular, was only the beginning of some potential clinical applications in this developing human story. There was much more to come.

What of the 'c' word – cancer?

That comes next.

17.

THE BIG ONE: CANCER

Within a few years after Dr. Bounous first discovered the quantitative immuno-enhancing property of his special whey protein concentrate, he and co-workers were daring enough to attempt, and surprisingly to prove, the anti-tumor effect at least in *mice* fed with a 20 gm whey protein / 100 gm diet(26). In particular, the chemically-induced colon tumors in these mice appeared to be similar to those found in humans insofar as the type of lesions and the chemotherapeutic response characteristics were concerned. Furthermore, the results with the whey protein concentrate feeding appeared to exert an inhibitory effect not only on the initiation of cancer, but also on the progression of tumors(61). Noteworthy, all the results were again initially observed in *mice.*

Similar results were also subsequently obtained in *rats* by Australian investigators. Dr. Graeme McIntosh and co-workers measured the input of different dietary protein sources on the incidence, burden and mass index of intestinal tumors induced by dimethylhydrazine in *rats*(27). They also found that whey proteins, in particular, offered considerable protection to the host against the chemically-induced tumors relative to other protein sources examined. Note that this was a different (second) species to show the same result, whatever the actual mechanism(s) involved.

Once again, this was reassuring confirmation for Dr. Bounous and his group. The dramatic anti-tumor effects of whey proteins were indeed real.

CANCER ORIGINS

Theories of the origin of cancer involve one or more of three kinds of molecular processes:

(i) free radical attack on key genes (oncogenes) in the DNA that cause on/off switches to initiate replication processes in renegade cell nuclei;

(ii) oxidative changes involving highly reactive oxygenated species that cause redox type phenomena which create havoc in cells; or

(iii) the poisoning effect of xenobiotics (foreign chemicals) that are toxic to normal cellular metabolism and biochemistry.

All these molecular effects are proposed to alter gene expression in one way or another. This leads then to mutations or changes inside the cells. If these conditions persist, the cells continue to malfunction and propagate the dire consequences.

Research on glutathione has elucidated the key role that this singular intracellular molecule plays in countering all three types of molecular processes just described above(62). Glutathione is the cells' principal free radical scavenger and antioxidant, as well as a key player in the enzyme detoxification of carcinogens by conjugation(35, 36).

The search for ways to inhibit cancer cells without injuring normal cells has been based over the years on an inadequate effort to identify the metabolic parameters in which cancer cells are at variance with normal cells. One such function could very well be the all-important synthesis of cellular glutathione(55). Indeed, researchers found elevated intracellular glutathione levels in their cancer cells compared to normal cells[62], presumably related to their proliferative activity. More specifically, the earlier *in vitro* assay developed by Dr. Baruchel, was used to show that Bounous's special whey protein concentrate caused *GSH depletion and inhibition of proliferation of cells in a rat model of human mammary carcinoma*[63].

The selectivity demonstrated by Baruchel in these experiments could be explained by the fact that GSH synthesis is negatively inhibited by its own synthesis. As mentioned earlier, since baseline intracellular GSH is higher in tumor cells, it is easier to reach the level at which negative feedback inhibition occurs in this cellular system, compared to a non-tumor cellular system.

In addition, patients taking Immunocal™ exhibited substantial increases in NK cell activity[64].

In any case, the dramatic anti-tumor effect that Dr. Bounous discovered in *mice* was crying out for an attempt at human application. Ever since that morning when Dr. Bounous first dissected the whey-protein-fed mice and witnessed with astonishment those pink livers, he just knew he was on to something even bigger than he had ever dreamed. This must be relevant for some suffering patient somewhere, even though it could never help his late mother.

But how could one make meaningful and responsible decisions for an early human trial?

PATIENT SELECTION

In testing a new approach to cancer treatment, patient selection and accrual are difficult issues. The selected patients must have a reasonable life expectancy, and must have measurable disease. They must not be denied available standard therapy and must, wherever possible, understand the rationale of the proposed treatment in order to be able to sign an informed consent. Thus a small number of subjects were selected for a pilot study and, for ethical reasons, a control group was not included.

A Phase I-II clinical trial was undertaken to test the effect of

Immunocal™ in five patients with metastatic breast cancer, one with pancreatic cancer and another with metastatic adenocarcinoma to the liver of unknown primary. Under the circumstances, it proves difficult to enlist patients for this kind of innovative trial. It is not surprising that the few consenting patients prove more often than not, to be quite seriously ill. The inclusion criteria for these seven select patients were as follows:

1. Histology (microscopic slides for cell type analysis and pathology) was available for review.
2. Measurable disease was present in soft tissues, lymph nodes, lungs, liver or bones.
3. No chemotherapy, radiotherapy or hormonal therapy had been given within three months of starting therapy with Immunocal™
4. Informed consent was obtained.
5. Life expectancy was at least six months.

For this small study, Drs. Bounous and Baruchel in Montreal collaborated with Drs. Renee Kennedy, George Konok and Timothy Lee at Dalhousie University and the Halifax Infirmary Hospital in Nova Scotia, Canada(65).

Each patient had a full clinical assessment at the time of entry into the study and at three and six months after initiation of therapy. Relevant imaging studies were done at the same time intervals. They were also seen bi-weekly for the assessment of their general condition, weight, tolerance to and compliance with the Immunocal™ therapy. Complete blood counts, serum albumin, total protein and standard liver function tests were obtained. An aliquot of blood was drawn for lymphocyte isolation and measurement of intracellular glutathione.

Immunocal™ was administered as a daily dose of 30 gm. of powder dissolved in a liquid of the patient's choice. The patient's diet was assessed for the six months to rule out excessive protein intake which has been shown to potentially enhance tumor growth. A brief description of the individual patient's history, clinical course, radiological and laboratory results follow below.

CANCER - CASE REPORTS

Patient (1)

This 37 year old woman received standard therapy for a Stage II carcinoma of the breast in 1990. Three years later she was investigated for increasing fatigue and shortness of breath. Chest X-rays showed a large right pleural effusion with a pathological fracture of the right eleventh rib. Cytology of the pleural fluid and a pleural biopsy were positive for metastatic breast cancer. In June 1994, when there was evidence of mild disease progression she was started on Immuncal™.

After starting Immunocal™ she reported an increased sense of well being and her weight remained stable at 61kg. After three months of therapy, her chest X-ray demonstrated significant clearing of the effusion, and evidence of healing of the pathological fracture of her eleventh rib. A corresponding bone scan showed the lesion to be less intense. After six months her clinical condition was stable, her chest X-ray essentially unchanged. A repeat bone scan suggested slight distal deterioration in her bone metastases.

The relevant laboratory data prove to be very interesting. Her pre-treatment haemoglobin was low and her platelet count abnormally high; abnormalities which are both seen with malignancy. As her treatment continued these values both came into normal range. The overall leukocyte count did not change but the lymphocyte count rose. Serum protein and albumin values remained stable, so it appears that the amelioration in the other parameters cannot be explained by nutritional factors alone. In agreement with her clinical improvement and the improvements in her basic blood data, this patient's GSH levels fell from abnormally high values at the start of the therapy to be within normal range.

Patient (2)

This 49 year old woman in 1986 underwent a left modified radical mastectomy for a Stage II invasive ductal carcinoma. She received 12 cycles of adjuvant chemotherapy. In 1990 she developed a persistent cough and a chest X-ray revealed two nodular densities in the left lung compatible with metastatic breast cancer. Her chest X-rays showed progressive disease and the patient refused to consider standard therapy.

In June 1994 she was started on Immunocal™, which she tolerated well. Her first set of chest X-rays taken three months after therapy showed an approximate 10% increase in the size of the lesions.

The subsequent X-ray in January 1995 showed no further interval change with apparent arrest of the progression of the pulmonary metastases. The patient remained completely asymptomatic with a stable weight at 49 kg.

In this patient the haemoglobin showed a minimal rise into normal range, while the leukocyte, platelet and lymphocyte counts which were normal at the initiation of therapy, remained so. The serum proteins remained singularly unchanged during the study.

Although this patient appeared to have a good clinical response, the GSH levels did not consistently fall. The levels however, even at their highest levels, were almost within normal range, so a significant drop could not really be expected.

Patient (3)

This 64 year old woman had an invasive ductal carcinoma resected from her breast in November 1989. The patient was extremely thin and there were no palpable lymph node abnormalities. At her request she did not undergo a staging axillary lymphadenectomy. Later, in June 1994, she presented with a palpable right supraclavicular lymph node 2 cm in diameter, which upon fine needle aspiration cytology was found to contain metastatic carcinoma. A routine metastatic assessment was otherwise negative.

She was started on Immunocal™ at the same time and after six months of therapy the metastatic lymph node decreased in size and measured one cm in diameter. A repeat fine needle aspiration remained positive for viable adenocarcinoma cells. Clinically she remained quite active and healthy. Her weight remained unchanged at 43 kg during the course of the study. She tolerated the Immunocal™ with no untoward side-effects.

Patient 3 was a clinical responder. Her haemoglobin reflected this, rising into normal range with treatment. The platelet and lymphocyte counts started off normal and remained so. Note again that the improvement cannot be explained by the nutritional effects of Immunocal™ as the serum protein levels did not reflect significant nutritional improvement.

The GSH levels fell from a significantly elevated level to within normal range, a trend that followed her clinical course as well as the changes in her haemoglobin.

Patient (4)

This 48 year old woman initially presented in September 1990 with a carcinoma of the left breast. A left modified radical mastectomy was performed for a Stage II carcinoma. She received adjuvant chemotherapy and at her request, Tamoxifen.

In February 1994, she experienced fatigue and on investigation was found to have widespread metastatic disease. She was started on Immunocal™ and initially, for the first eight to twelve weeks she had an excellent clinical response with a remarkable feeling of well-being. She was able to take a physically demanding vacation. After this initial period of improvement, her condition gradually deteriorated with the development of diffuse musculoskeletal pain. Her initial GSH was high, then it fell with therapy but thereafter followed a rising trend.

Patient (5)

This 73 year old woman presented in June 1994 with obstructive jaundice secondary to a carcinoma in the head of the pancreas. At laparotomy it was found to be unresectable and a palliative choledochojejunostomy and gastric-jejunostomy were performed. She had an uneventful post operative course and, at her request she was started on Immunocal™ in July 1994.

The CT scan of the abdomen at three months of treatment showed stabilization of her disease. At six months, she was feeling slightly tired, having just recovered from a flu-like illness, but her CT scan showed evidence of progressive disease with an increased tumor size, and the development of mild ascites.

After an initial response with an increased sense of well being, clinically she started to deteriorate and this deterioration was associated with a steady increase in her lymphocyte GSH levels and a progressive fall in peripheral blood lymphocyte count.

Patient (6)

This 72 year old woman in 1985 had a modified radical mastectomy for a Stage I invasive ductal cancer. In 1989 she developed metastatic bone disease and received radiotherapy as well as chemotherapy with CMF for six cycles. When her disease progressed she was offered only supportive care and pain relief as she had refused second line chemotherapy.

In July 1994 she was offered Immunocal™. Initially, she did very well with increased strength and energy, becoming noticeably more mobile. This period lasted about two and a half months after which she became progressively more fatigued and inactive. Her imaging studies showed at that time, progressive disease.

The GSH levels were elevated at the start of therapy, and there was a brief decline to normal range, but as with her overall status, this did not remain so and they rose to even higher than pretreatment values as her disease progressed. Her blood haemoglobin and lymphocyte count progressively declined.

Patient (7)

This 76 year old woman presented with abdominal pain. On

investigation she was found to have biopsy proven metastatic adenocarcinoma of the liver. A primary site was never found. As there were no specific treatment options open to this patient she was started on Immunocal™ in August 1994. Her clinical course remained stable until January 1995 when she developed increased nausea and vomiting. She was admitted to hospital where her medications were readjusted and six months later, she was once again thriving clinically and had almost regained her usual weight of 74 kg, after having lost 4 kg. Her blood profile was stable and her abdominal CT scan showed minimal progression of her liver lesion. Her GSH levels showed an initial fall with a slow subsequent rise probably heralding an impending clinical progression of her disease. Blood haemoglobin and lymphocyte count remained stable within normal range.

What should be made of these few preliminary case studies?

An interesting phenomenon was observed in that every patient experienced a period of improved sense of well-being, which while difficult to quantify, was appreciable in all patients. In some, it led to a perceived improvement in the quality of life for at least a short duration, and some patients were able to perform activities they could not previously. The number in the study was too small to draw firm conclusions from this, especially without matched controls, but ***the results were encouraging***.

Most of the patients could be characterized as "responders", or "non-responders" based on the correlation between their clinical course and their blood GSH levels. Six patients started with substantially elevated values compared to known normal values as well as a simultaneously performed normal control. Those who showed only a brief response had an initial drop in their lymphocyte GSH levels which corresponded to their clinical picture. However, as their disease progressed, the GSH levels began to rise, once again (patients 4-6). Those who had a favourable, and more protracted clinical response to Immunocal™, had noticeable and sustained drops in their GSH levels (patients 1,3). In some patients, positive clinical results were associated with increased haemoglobin (patients1-3) and peripheral blood lymphocytes (patients 1,3) and normalization of platelet counts (patients 1,5), all of which are indirect measures of disease regression.

ADJUVANT THERAPY

A major problem in the use of chemotherapeutic agents in cancer treatment is the protection offered by the defense mechanisms of the tumor cells. An important element of protection is represented by GSH which is an effective detoxification agent, relatively abundant in tumor cells. Immunocal™ favorably influences the GSH synthesis in normal cells. Hence, this nutritional supplement while exerting an inhibitory effect per se on cancer cell GSH and replication, could also be viewed as an adjunct to chemotherapy(65). Lower GSH levels could in fact, render cancer cells more vulnerable to chemotherapeutic agents. The validity of this assumption is substantiated by the observation by Hercbergs and coworkers., that cancer patients are more likely to respond to chemotherapy if their erythrocyte GSH, and by inference tumor GSH, concentrations are low(66).

Given the technical and ethical difficulties in monitoring tumor tissue GSH during cancer treatment, peripheral blood lympocyte GSH levels were taken in the small study just described, as a reflection of those in tumor cells. This assumption is substantiated by a previous study of a large series of cancer patients in whom erythrocyte GSH concentration was found to reflect tumor cell GSH on the basis of differential tumor responses to chemotherapy(66).

In view of the advantages of reduced intracellular levels of GSH in tumor cells and increased levels in the host, it seems noteworthy to have found in the patients who performed well on Immunocal™ (#1, #3), the highest levels of initial lymphocyte GSH. These values fell into normal range soon after Immunocal™ was initiated and remained so through the six months of treatment. In patients exhibiting disease progression the GSH levels tended to rise. It is felt that the high lymphocyte levels of GSH occur as a result of a leaching effect from the tumor to the lymphocytes. These lymphocyte values are therefore felt to be an indirect measure of tumor levels, as were the erythrocytes in the previously quoted study by Hercbergs and co-workers(66).

This preliminary study, indicated at least that *this newly discovered property of whey proteins may be a promising adjunct in the nutritional management of cancer patients about to undergo chemotherapy.* Selective depletion of tumor GSH may in fact render the malignant cells more vulnerable to the action of chemotherapeutic agents.

So where was all this leading? First there were the positive effects of Dr. Bounous' whey protein concentrate in HIV-seropositive individuals, suggesting some possible and perhaps complementary role for such dietary invention. It was unlikely to be an alternative to the standard antiretroviral drug therapy. Then came the effective glutathione depletion in tumor cells and the inhibition of proliferation in a few selected advanced cancer patients, even with metastatic disease. This clearly was pointing to a need for further investigations but even from these early results, the prospect of modulating intra-cellular glutathione safely, effectively and conveniently was creating some excitement among some cancer researchers even though caution was still well advised.

The early clinical applications for Immunocal™ - this unique whey protein concentrate that Dr. Bounous had stumbled upon - were therefore very promising for some kind of therapeutic applications. Perhaps there was the genesis of some new '*drug*' on the horizon. But that could hardly be. The product itself was nothing more than a concentrated derivative from a common '*food*' staple in the typical diet in most parts of the world.

Could Immunocal™ be at the crossroads? 'Drug' or 'food'? Which way might it be taken?

18.

IMMUNOTEC RESEARCH, LTD

Immunocal™ is a **food**, not a drug," Dr. Bounous is always quick to explain. "Never mind that we manufactured it to the highest pharmaceutical standards: the drug companies we approached simply would not touch it. They think differently, they are structured differently and they operate differently."

'FOOD' OR 'DRUG' ?

The standard pattern for drug companies is to spend lots of time, expertise and money on research and development in their areas of interest, to find a new or modified pharmaceutical preparation with particular pharmacological indications. Then after tedious clinical trials and lots of government red tape, there is hopefully approval for release of the new preparation with some limited patent protection. The drug companies typically choose to initially recover the exorbitant development cost. Cheap preparations can then become high-priced prescriptions across the drugstore counters as the lay public consumes what they do not generally understand, to relieve whatever physical or emotional condition they cannot stand.

Of course, the challenge to gain market share is a purely marketing one, and since doctors are the gatekeepers, the drug oligarchy does a hard-sell job on the profession and a soft but highly visible promotion to the public. And most drug companies are very good at this. Read any doctors' magazine or watch television for an evening and the evidence is overwhelming. Profit margins therefore tend to be debatably high, but the public is vulnerable and sometimes desperate for any hope of relief.

Where could Immunocal™ fit in? It was discovered by a serendipitous breakthrough and had a common underlying mode of action by modulating glutathione synthesis with dramatic consequences. And it did this with no downsides. After all, it was akin to mother's breast milk proteins, so what could be more safe, especially with an intrinsic control by feedback inhibition to limit intracellular glutathione levels? Yes, it was proven safe and effective but it was also apparently most common in origin. It was hardly shrouded in mystery.

Despite dozens of well-conducted medical investigations and a mounting stack of published proofs of efficacy and safety, one overriding fact remained: Immunocal™ is derived from *milk*. It is a pure whey protein concentrate, necessarily manufactured with exquisite care to avoid conditions that would otherwise denature its precious, life-enhancing bioactivity. It was proven to be unique.

As a product, Immunocal™ had been discovered and substantiated in the mainstream of scientific research in a world-class health-science facility. It was not found by some little-known, self-styled 'doctor' with uncertain credentials, who would make some unsubstantiated claims based on private anecdotal experiences in some obscure laboratory or clinic. As we have seen, Dr. Bounous had been recognized by his peers as an outstanding medical researcher. His findings had been corroborated and published in collaboration with several other distinguished scientists in reputable peer-reviewed journals. Independent research in Japan and Australia had clearly produced similar results and no one had challenged the fundamentals. This was good, mainstream, fundamental medical research. Only such authoritative sources made a breakthrough of this magnitude credible, especially when the source of the invention is so common-place.

COMMERCIAL PREPARATION

Both Dr. Bounous and Dieter Beer were therefore adamant about the need for stringent quality control in any scale-up for commercial application. They began their search for a supplier of such specialized bioactive whey protein concentrate with excessive demands. Only a very large, reputable dairy would have the flexibility, the resources and the expertise necessary to deliver a reliable product on a consistent basis. They would have to be again in the mainstream. There would be no compromise.

Their search led them to the dairy heartland of America. In one state alone there are 284,000 cows, most of whom (greater than 98%, in fact) are Holsteins. They are milked on a 305 day lactation cycle during which each cow will produce approximately 19,000 pounds of milk, or over 80 pounds of milk per cow per day. Most dairy herds are milked twice a day in traditional herringbone style milking sheds. Some of the new milking sheds are the rotary type which allow one person to milk up to 200 cows per hour. These cows are fed at a rate of about 60 pounds of feed per cow per day in order to maintain this production. About 50% of their feed consists of dry hay and alfalfa and the balance is made up of nutritious grains such as corn and sow. Some farmers feed their cows corn silage.

A full range of tests is carried out prior to milk reception at the plant including tests for Betalactam inhibitors (first and second generation penicillin antibiotics), Amoxicillin, Ampicillin, Ceftiofur, Cephapirin, Gloxacillin, Hetacillin and Penicillin). There is also random inhibitor screening for sulfonamides, macrolides, aminoglycosides, and tetracyclines. In addition, the FDA (food and drug inspectors) perform random screening for all families of drug residues on individual milk suppliers.

The proprietary process involves a series of treatments to isolate the specialized undenatured bioactive proteins. In the initial step, cold pasteurization (or microfiltration with large pore ceramic filters,) removes gross bacteria mould and dirt contaminants. This is followed in a second step by a low temperature (72 - 73 °) pasteurization for about 16 seconds to remove residual bacteria and other pathogens. After precipitation of curd (cheese), the supernatant whey is treated by a slow ultrafiltration process through a fine pore filter. The 'permeate' which passes through includes lactose and any water-soluble toxins and pesticides. The fat soluble ingredients are removed early in the preparation, so there is no rancidity and less risk of inadvertent poison or allergy. The 'retentate ' is a unique protein mixture with 20% solids. Through several rinses (diafiltration) lactose is removed and by the same mechanism (process), most of the water-soluble and undesirable materials are eliminated at the same time. It is then spray dried by a lenient technique to avoid denaturation and yield a fine white powder. Flow characteristics can be improved by the addition of up to 0.7% food grade lecithin. Bacterial counts are monitored routinely. The typical accepted standards go as high as 10,000 bacterial counts per gram with no pathogen, whereas Immunocal™ is consistently around 400 – 500 counts per gram. The finished product is packaged under pharmaceutical-grade conditions.

Indeed, so carefully and specifically was this "food product" prepared that it almost became a "humanized native milk serum isolate". It was aimed at preserving, in their natural form, the specific cows' milk proteins which share with the predominant proteins in human milk, the same rare GSH-promoting components. Naturally, the special immune protective properties of mothers' breast milk are well known(67, 71). Breast-feeding is known to be superior to cow's milk-based formulas of similar nutritional efficiency with regard to health of human babies. Among other things, it protects against infection and the incidence of several types of childhood cancers, especially with long-term breast-feeders.

PHARMACEUTICAL COMPANIES ELECT TO PASS

Despite the medical breakthrough Immunocal™ represented, government regulations in North America required it to be labeled as food. It could be sold over the counter without any necessary prescription or even directly by person to person, without license or restriction. It was safe and virtually harmless, as much as any other usual food or food product. Its benefits, therefore, should be far more accessible and automatically cost the consumer far less than a typical pharmaceutical product, even if a drug could be devised that would offer the same proven and potential human health enhancements.

Startling and effective though Immunocal™ might be, major pharmaceutical companies weren't interested in a product which had such familiar origins and which offered them such low profits. How could they make it something more than what it truly was? After all, it was still a *milk* derivative. A unique product, yes, but from a ubiquitous source. And unlike most drugs, it did cost relatively little to make the initial research find. But then, it cost lots to make the final commercial product, on an on-going basis. Most drugs in contrast, usually have high expenditures in research and development but comparatively less in final preparation or manufacturing. Immunocal™ lacked mystery and therefore could not bear exorbitant mark-up, over and above the cost of goods. So North American drug companies elected to pass on it.

That fact certainly did not shock Dr. Bounous. After all, how many years had he tried in vain to convince his medical colleagues to try proven 'elemental diet' nutrition first, in certain hospitalized patients? Yet the professional gatekeepers preferred the stronger, costlier intravenous measures, with their usual side effects, even when the simpler means of enteral nutrition had been proven to be effective(13,14,72).

For the first time, in 1967, Bounous had demonstrated that common nutrients may protect or heal the intestinal mucosa by virtue of the particular form and mode in which they are delivered to the intestinal lumen and their availability to the mucosal cells. The definition of these elemental diets as 'medical food' was justified by the fact that enteroprotection is associated with, but independent of, systemic nutrition.

In the two decades, following 1967, Bounous emphasized the concept that enteral nutrition should be, whenever possible, the ideal treatment for severely injured patients.

Nutrition first, where appropriate: drugs second. That might seem the sensible, logical and conservative approach, but things do not yet work that way. In the United States, in fact, pharmaceutical companies in 1995 led all other industries in profitability. So-called 'alternative approaches to health,' very often are far less risky and usually much less costly than prescription drugs, but they are often adopted only when the public cries out or when all else fails.

THE HEALTH FOOD CHALLENGE

But the typical natural supplements that one can buy in a health food store or through direct distribution to the consumer are, by government edict, labeled as 'food.' Therefore, no unapproved claims for possible healing or therapeutic benefits are legally permitted, and the interested consumer must learn for himself how to utilize such products to foster his own health condition. As a result, the food supplement industry is flooded with much misinformation and quackery. It would almost appear that the responsible supplier of a well-researched, proven, high-quality food supplement, who markets an effective product like Immunocal™ consistent with the government regulations, is at a major competitive disadvantage (in the short term only, perhaps). They must operate in the same marketplace with reckless and irresponsible profiteering entrepreneurs, who exploit the ignorance and vulnerability of a lay public with exaggerated claims for poor quality products that have little scientific support.

Contrast this with Asia or Europe, and especially Germany, where the scientific and governmental attitudes towards botanical products and certain other natural remedies are altogether more accepting and cooperative. In Germany, for example, a government body similar to the U.S. Food and Drug Administration tests and approves botanical medicines. Natural substances sold there as "nutritional remedies", or "alternative medicines" or "health promotion products" are manufactured to pharmaceutical standards and clearly labeled as to their suggested doses, common uses, possible side effects and known contraindications.

Obviously Germany tries to create no artificial differences between the expected standards for pharmaceuticals and natural health remedies. All are recommended to be scientifically investigated and tested whenever possible, before being clearly labeled and then prescribed or sold over the counter. All claims for each regulated product, whether natural or synthetic, should be equally well researched and backed by government screening and tests. In the absence of acute or critical conditions, the consumer reserves the right to choose what type of prescribed drug therapy or alternative 'natural remedy' they consider appropriate for themselves. If the public is informed, they can be left to make their own responsible decisions. This remains a turf war in North America.

Realistically, Dieter Beer and Gustavo Bounous knew they would encounter a plethora of differing governmental regulations throughout the world if and when they commenced to market Immunocal™. Ironically, here was one of the world's purest, and most natural health products, they believed, one as universally efficacious as mother's milk, totally harmless and with a range of powerful, proven immune system benefits ... yet it could not be labeled as such in North America. Indeed, government regulations absolutely prohibited them from making any medical claims whatever for Immunocal™. What an apparent contradiction to all the popular exaggerated marketing hype, advertised and broadcast day after day across the continent. What you see is seldom what you know, much less what you get. It is a farce in the public's face. But that's the way it is. Can the responsible guys ever win in that marketplace? Perhaps so.

A MARKETING SOLUTION

How, then, does one market and distribute so authentic, so sophisticated and important a product as Immunocal™?

If any person could solve that problem, it surely would be Dieter Beer. He has the uncommon ability to focus with intensity, to think creatively and to derive practical solutions for stubborn problems. It is a gift cultivated by persistent effort and experience. Beer's brilliant, encyclopedic knowledge of international business, marketing and entrepreneurship stemmed from years of hands-on experience and well-proven success. He is a visionary and a problem-solver par excellence, equally well versed in matters of business, law and the aspects of corporate growth management.

This time, however, Beer admitted that he was stumped. If one could not fully describe, advertise or otherwise claim even well-proven and documented facts about a product's efficacy, how does one position it in the market place? If pharmaceutical companies would not distribute it, through their monopolized channels, Beer and Bounous reasoned, perhaps the alternative outlet could be through health food stores. But then, must Immunocal™ languish on health food store shelves, indistinguishable from the hundreds of other natural products with so much less scientific rationale and dubious quality control, and which ultimately become known to all-too-few of those consumers who need them? How would it be advertised or promoted to encourage consumer use? After all, Immunocal™ was a product of mainstream health science, it was not some backyard discovery or amateur concoction. It was science at its best, exploiting the benefits of nature's food supply for its most fundamental protective value.

Dieter Beer strongly believed that everyone needed to experience the benefits of Immunocal™ and deserved a chance to have them. He was eagerly looking for a way to communicate the value of this breakthrough product to the wider population and then for some convenient means to get it to them. This was a personal mission that he had undertaken and that he had to fulfil.

His solution came from a most unexpected source. The Medical Director of a large pharmaceutical firm which had shown some early interest in Immunocal™, had concluded that as fine a product as it was, it would not make any commercial sense for his company; in fact, or for any drug company, he thought. It was that doctor who first steered Dieter Beer down a totally different road one day. He suggested to him that the answer to his onerous dilemma had narrowed into one final option, the only one which seemed at all feasible: Network Marketing. That idea seemed light years away from every other business concept Beer had embraced in the past. Had the source of that suggestion been any less formidable or credible, it may never have connected.

"I knew very little about this method of marketing and distribution," Beer admits, "and at first I was very prejudiced against it. It somehow brought the idea of those old illegal pyramid schemes to my mind ... I was innovative yes, but still I was a conservative businessman."

Typically open-minded, thorough and objective in his approach, however, Mr. Beer investigated this network marketing business and studied a few existing business plans. When he was ultimately satisfied, he decided that such a method represented the best way to proceed to promote and distribute Immunocal™. He admits that he initially felt less than fully convinced: his remaining doubts, however, soon would vanish.

NETWORK MARKETING

What did Dieter Beer learn about Network Marketing that dispelled his early apprehensions?

Network Marketing is perhaps one of the biggest kept secrets in business in North America today. A quiet revolution is taking place in the way new products and services are being introduced to the market place. This is particularly true of some innovative, high-quality, effective products that evoke loyalty among satisfied customers. Healthcare products especially are ideal for this method of distribution in an era when wellness, fitness and lifestyle are potent market drivers.

Networking is a respectable methodology today, attracting the support and usage of the corporate world including major Fortune 500 companies. It is a worldwide phenomenon with an estimated 25 million participants and over $100 billion in annual sales. These numbers continue to climb. As a very legitimate business, the network or multi-level method is now being taught in academic schools for its proven effectiveness in fast start-up operations and for the remarkable customer satisfaction due to its high levels of personal service. As a cornerstone of the New Economy, it is leading the trend toward home-based small businesses with easy access through in-line computer technology.

Network marketing is a contemporary career opportunity – attracting professionals, entrepreneurs, baby boomers and people from all walks of life. It is certainly a lucrative industry for the successful pacesetters among these would-be, small-business opportunity-seekers. Network marketing is making more millionaires in record time because of their practical leverage of time and energy, coupled with not so uncommon exponential growth.

But this is a relatively level playing field where one's resumé is always right, since they can always begin where they are and grow from there. For those who do, it is a fulfilling experience. It legitimizes the human factor in commerce, by enhancing relationships and converting the human personality and other personal qualities into real business attributes.

And the end is not yet in sight. There seems to be an unlimited horizon for this 'megatrend' as people continue to 'cocoon' and network to advantage. The futurists are predicting a so-called 'Wave 3' tidal influence on the way we all live and work.

Doubts and prejudice aside, Beer soon realized that network marketing could do far more than simply offset any negative effects of regulatory restrictions and the perceived lack of over-the-counter sales.

Perhaps the greatest value he acknowledged would be the power of direct communication from person-to-person. This would be an honest, caring and pointed exchange of real information tailored to each customer's needs, and not some subliminal suggestion or other advertising hype. Real people could spread the true word. Through appropriate channels, consumers could receive pointed and reliable answers to their immediate needs and questions. In a word, it could be a caring, communication network. What better way to communicate such a valuable message of health and hope.

Network Marketing could also spawn some other very significant benefits for everyone: from those at the corporate level, to a growing cadre of individual entrepreneurs, and eventually to countless thousands if not millions of their clientele. Everyone could win. Dieter Beer knew from experience the value of opportunity and the joy of success. He would take special pleasure in seeing other aspiring entrepreneurs, no matter the scale of their business, assume control of their life's agendas. They could reassess their priorities, change direction and focus, and invest time and effort to develop their latent capabilities en route to new levels of accomplishment and success. That was all part of the wonderful magic of network marketing. This could be not only a breakthrough product introduction but a breakthrough opportunity offering.

THE MERGER

Beer's crash course in network marketing began the day he met one Charles H. Roberts, a Montreal entrepreneur, family man and sports enthusiast. Roberts would persuade him to establish a new company to capitalize on the Immunocal™ discovery. They would agree to call it **Immunotec Research, Ltd.**, and Roberts would be named its President.

Roberts, whom everyone calls "Chuck", has an open face which readily forms an unprententious smile, so characteristic of his warm endearing personality. He is tall with the long-legged stride and high energy of someone more at home outdoors than behind a desk, and a mind apparently accustomed to handling multiple problems and decisions simultaneously. He is gutsy and spirited and obviously given to seeing the big picture and the long range view, refusing to be distracted by pettiness or detail. He tends to be very straightforward and practical.

Chuck met Dieter Beer through his very good friend Alex Konigsberg, an internationally recognized franchise attorney. Konigsberg, who is as quietly charming as Beer, with the same courtly qualities of cordiality and politeness, learnt of Beer's marketing challenge through the patent lawyers in his firm. He obviously accomplished a coup when he brought the two men to the negotiating table.

"Dieter showed me all the research results and publications that first day," Roberts recalled. "We called the patent lawyers and I saw all the patents awarded at that time. I immediately saw the potential for this new product. Even when I learned that at the time, the projected cost for a month's supply of Immunocal was costing out at two hundred dollars a box and even so, it only had a four-week shelf life, I still wanted to sign on."

In some ways Roberts' business career in Canada paralleled that of Dieter Beer's in Europe. Both men are self-made entrepreneurs with several successful business start-ups to their credit. Each had engaged in international marketing and sales, networking globally, but Roberts' career included one element Beer's lacked: multi-level marketing. There's an amusing and serendipitous story behind that.

Chuck Roberts began his business career at Kraft Foods Ltd., an international company that his grandfather was instrumental in bringing to Canada along with others and served as its "founding" president. In 1964 he joined Pedigree Canada as sales manager, rapidly earning the position of General Manager, and in 1972 he purchased the company. He was on track for more success. Chuck soon acquired warehouses and branches throughout the United States and enlarged his business to establish both the Pedigree and Jean-Claude

Killy skiwear lines, which he also owned.

Later he inaugurated a new line, Descente Canada, and supplied sports gear to American, Canadian, Swiss and Spanish Olympic ski teams. Descente Canada skiwear has maintained its position as the industry's top-rated brand.

After Chuck elected to take early semi-retirement, he sold most of his sports businesses and took up more avid boating through the east-cost intercoastal waterways in his forty-seven foot yacht. He was then doing what he wanted to, but he was somewhat restless and still enterprising. His brother-in-law would soon entice him to assume a new challenge for himself and not long after, he was marketing international franchises for a multi-level company that his relative was running. He served his time doing that with global success and then retreated yet again to the life of leisure.

As their three daughters and one son approached adulthood, Chuck's wife Suzanne Roberts decided to become involved in a new and different venture for herself. She wanted a challenge and saw the fulfilling prospect of working with people which she always enjoyed. She became a distributor within a Canadian multi-level sales company, and discovered she had a natural affinity for the business. Tall, blond and vivacious, with energy and enthusiasm apparently equal to that of her husband's, Suzanne soon achieved success in her business. For the first time Chuck Roberts took a good look at the way direct marketing works at ground level and soon wanted to try it himself.

The result was Evergreen Products, Ltd., a company Chuck founded to distribute herbal preparations and other natural products in Canada and into the United States.

"Dieter was a financier who never knew or understood the direct marketing business," Roberts said. "I had sold several international franchises, including Brazil, all of Europe and the Eastern bloc, South Africa, Indonesia and the Middle East. I understood what we were getting into."

Some strategic alliances or business mergers seem to coalesce naturally and with great ease. Here was one of them. Dieter

Beer and Chuck Roberts came together after clean and relatively easy negotiations.

"I remember Dieter and me leaving all the lawyers inside, going out together for a chat, away from the laborious paper work and number crunching. We were getting bogged-down in technicalities when we both realized we had a job to do, a mission to accomplish. So we mutually committed to resolving all the issues, shook hands and started to get on with the job."

That was it. Immunotec Research Corporation which was still essentially Dieter Beer and Dr. Gustavo Bounous, would merge with Evergreen Products Ltd to form a new company to be called simply **Immunotec Research, Ltd.** That was a perfect self-descriptive name. It kept the emphasis on research and on the immune system, just where the principals wanted it.

"We signed the deal in November, 1996 and opened our doors on February 1, 1997," Mr. Roberts said. "There are always many pitfalls in building this kind of business. You organize before you sell, and we made sure we were well structured before we opened the doors."

Dieter Beer had funded the research on Immunocal™ and Chuck Roberts would fund the new company start-up. Those days he and Beer often worked late into the night as they decided how to structure the company and chipped away at refining the product. Soon they were able to lower production costs and cut the consumer's cost for Immunocal™ in half. On the scientific side, the product's shelf life was extended under Dr. Bounous's guidance from just four weeks to *one full year*.

"We say optimum shelf life is one year," Chuck Roberts explained. "But we know the real stability under normal conditions is better than that."

"I like the product's packaging, which features an antibody logo," he would say. "My grandfather created Velveeta cheese and we're still using the same Velveeta packaging today like he used two generations ago. That's possible if you get it right the first time."

With her husband Chuck at Immunotec's helm, Suzanne at first handled corporate public relations and trained the company's new sales associates. As the organization grew during that fast-paced first year, she gradually relinquished most of her other duties to concentrate on sales training, for that is really her forté. Presently the company's only female partner, Suzanne holds a distinct vision of how its sales and marketing should be shaped.

"We represent a very serious product, and we want to maintain a high level of dignity and respect within our industry," she said. "We'll never be one of those rah-rah companies. We're simply going to help each of our independent associates to accept the wonderful business opportunity Immunotec Research offers, to work with conviction and real enthusiasm, and to hopefully become financially independent.

"Dr. Bounous has created something truly amazing." she added. "The rest of us want to do our part to help get it out to the world."

THE FITNESS FIT

Chuck brought another key player with him from Evergreen. It was his friend and associate, John Molson. He has a completely different story but shines as a star in his own right.

John H. Molson, now Vice President of Immunotec Research, Ltd., spent two years in corporate lending with the Royal Bank of Canada. Expert in credit risk analysis, he brings not only a broad background in finance to the new company, but another, far more surprising perspective as well.

Molson is a well-known, highly trained athlete who for years has regularly and successfully competed in some of Canada's major sporting events. In 1994 he won both the road race and the individual time trial at the Quebec Cycling Championships. His team later that year aced the Quebec Team Triathlon Championships.

Intense about personal health and fitness, and knowledgeable about the physical costs of human sports competitions, John brings to Immunotec Research an unusual breadth of varied business experience combined with specific knowledge of the trained athlete's daily physical challenges and the necessary discipline. His competitive sports background promises to guide the company into some interesting future experiments in measuring Immunocal's restorative effects on the bodies of elite athletes. He is actively engaged in facilitating on-going research interests that continue to escalate as more and more potential implications and applications for the breakthrough product are recognized in wide-ranging disciplines. As active liaison, he tries to manage the Bounous connection in this exploding research thrust. There is more to come.

The Molson family long has been known throughout Canada for two things – beer and philanthropy. John's grandfather founded the famous Molson Brewery, where John worked in International sales and marketing. His grandmother for years fed disadvantaged children; today her grandson occupies himself with various causes such as Homes For The Homeless. To Immunotec, he adds a very human touch.

Both Chuck Roberts and John Molson possess firsthand experience with the ongoing challenges inherent in producing safe, high-quality food products. The men's separate backgrounds in their respective industries contribute significantly to the continued excellent quality of Dr. Bounous's milk whey concentrate product.

PATENT PROTECTION

Immunotec Research Ltd had a number of important strengths that would provide more than adequate foundations for future growth: stable, sustainable growth. The key principals were committed to their mission for the long term and with their proven track records and their good chemistry, nothing seemed unattainable. The product Immunocal™ had proven to be a major breakthrough in health and disease and with such a consumable base, there were exceptional prospects for success.

But early in the game, Dieter Beer recognized the importance of protecting the major discovery and its applications from would-be imitators and those who might try to capitalize unworthily. So he spared neither effort nor cost to gain full patent protection for Immunocal™ and at the time of writing, a number of method of use patents have been issued. Their exclusive claims ranged from being an effective biologically active food supplement, to enhancing cellular glutathione to improve immune response, to treating HIV-seropositive individuals and being an anti-cancer therapy. These patents have been issued in different countries including the US, Canada and Australia. For completeness, the abstracts are detailed next.

U.S. Patent 5,230,902
Date of Patent: July 27, 1993
UNDENATURED WHEY PROTEIN CONCENTRATE TO IMPROVE ACTIVE SYSTEMIC HUMORAL IMMUNE RESPONSE

This invention provides a method of improving the humoral immune response or increasing the concentration levels of glutathione in mammals, which comprises administering orally to a mammal a therapeutically or a prophylactically effective amount of undenatured whey protein concentrate which has a biological activity based on the overall amino acid and associated small peptides pattern resulting from the contribution of all its protein components. A method for improving the humoral immune response in mammals also is disclosed which comprises administering orally to a mammal the combination of a vitamin supplement containg vitamin B_2 in an amount in excess of minimum daily requirements and an effective amount of undenatured whey protein concentrate. This invention further provides a dietary supplement for a mammal which comprises an effective amount of Vitamin B_1, and B_2 and a therapeutically or prophylactically effective amount of whey protein supplement.

U.S. Patent 5,290,571
Date of Patent: March 1, 1994
BIOLOGICALLY ACTIVE WHEY PROTEIN CONCENTRATE

The present invention is concerned with a whey protein composition comprising a suitable concentration of whey protein concentrate wherein the whey protein concentrate contains proteins which are present in an essentially undenatured state and wherein the biological activity of the whey protein concentrate is dependent on the overall amino acid and small peptides pattern resulting from the contribution of all its proteins components and a method of producing said whey protein composition. The invention also relates to several applications of said composition.

U.S. Patent 5,451,412
Date of Patent: September 19, 1995
BIOLOGICALLY ACTIVE UNDENATURED WHEY PROTEIN CONCENTRATE AS FOOD SUPPLEMENT

A method of increasing the rate of synthesis, rate of replenishment, and concentration levels of glutathione in animal organs, comprising the step of administering to an animal a therapeutically, or a prophylactically effective amount of undenatured whey protein concentrate containing substantially all the heat labile whey protein present in milk.

U.S. Patent 5,456,924
Date of Patent: Oct. 10, 1995
METHOD OF TREATMENT OF HIV-SEROPOSITIVE INDIVIDUALS WITH DIETARY WHEY PROTEINS

A method of treating HIV-seropositive individuals so as to increase their blood mononuclear cell glutathione concentration and to maintain or increase body weight which comprises administering to HIV-seropositive individuals a substantially undenatured whey protein concentrate, wherein the substantially undenatured whey protein concentrate comprises substantially all the heat labile whey protein contained in raw milk, in an amount effective to increase their blood mononuclear cell glutathione concentration and maintain or increase body weight.

Canadian Patent 1338682
Date of Patent: Oct. 29, 1996
BIOLOGICALLY ACTIVE UNDENATURED WHEY PROTEIN CONCENTRATE AS FOOD SUPPLEMENT

... A whey protein concentrate comprising cystine to act as a precursor for glutathione synthesis and including heat labile protein components providing an amino acid and associated small peptide pattern adapted significantly to enhance the bioavailability of the cysteine thereby raising cell glutathione content and improving humoral immune response ...

Australian Patent 38812-93
Date of Patent: September 20, 1994
METHOD OF TREATMENT OF HIV – SEROPOSITIVE INDIVIDUALS WITH DIETARY WHEY PROTEIN

Australian Patent 46937-93
Date of Patent: January 19, 1995
ANTI-CANCER THERAPEUTIC COMPOSITION CONTAINING WHEY PROTEIN CONCENTRATE

Some thirty years after Dr. Bounous began his pioneer explorations into the role of nutrition in basic human functions, his crowning discovery – Immunocal™ – at last was to be put to full scale practical use.

Year after year of apparently slow and lonely efforts finally had culminated in a swift-moving, power-packed business entity. Immunotec Research, Ltd was clearly a different company looking into the future with a real scientific breakthrough in hand. After all, *the challenge to the immune system was now front and center in the health and wellness field, and here was the most dramatic new offering for self-help in over half a century.*

Would success arrive at last? Again, Dr. Gustavo Bounous stood, figuratively, on emotional tiptoe.

1998

"The first wealth is health."

- Emerson

19.

NEW HORIZONS

By early 1998, Immunotec Research, Ltd had grown in its first year at an astonishing rate. Almost ten thousand individuals were already distributing Immunocal™ throughout the United States and Canada. These entrepreneurs/distributors not only included small business types, salespeople, physicians and other professionals, but then a broad spectrum of other ordinary health-minded individuals had also become involved. They all had two things in common: (i) A knowledge and conviction of the major breakthrough that Immunocal™ represented for health and disease at the turn of the century, and (ii) a passion and commitment to spread the word and share its benefits widely, while at the same time, securing a brighter future as they capitalized on this unique networking opportunity.

One of these entrepreneurs was an opera singer who understood that Immunocal™ could reinforce his immune system and perhaps that of his colleagues during the demanding and strenuous opera season. Another was a professional tennis player who learned of its value in reducing oxidative stress post exercise and shortening recovery time. Still another was a driven Type-A realtor who had been burning the candle at both ends and was now feeling exhausted and depressed. People from all walks of life were finding out about Immunocal™ by tailored one-on-one communication as the network grew.

When asked the question, 'Who will this product help?' Dr. Patricia Kongshavn, a career specialist in immunology, went straight to the point.

"Number one, normal people," she said. "We're all likely to benefit from increasing glutathione. The normal exposure or wear and tear of daily living demands better protection. The average person's immune system slowly wears down, and to boost it again typically produces better body resilience, and such benefits as perhaps fewer colds and other common infections that are far too much a part of everyday life.

"Who should take it?", she continued, "Anyone who feels their immune system is down or in need of a boost for some reason or other. It is better to ask the opposite question: 'Who should not take it?' Or, 'who would not benefit?' "

Here then was a way to optimize one's immune system. The idea filtered out quickly. Blue-collar workers, athletes, homemakers, school teachers, police officers, health professionals, business people ... all these and more began to catch the vision. Immunotec Research, Ltd hardly recruited at all. Men and women interested in improving or maintaining their own good health and that of their families, entrepreneurs who wanted to seize the novel opportunity to spread the word ... they began increasingly to seek out Chuck and Suzanne Roberts and others in the company. The news of the breakthrough was spreading and the small start-up company was quickly gaining momentum.

DIRECTOR OF RESEARCH

Sales figures climbed as word of the company's exceptional Direct-Tech™ compensation plan attracted increasing numbers of new networkers and distributors. These activities obliged Dr. Bounous, still Immunotec's Director of Research and Development, to work steadily, with John Molson's able assistance, at answering queries from physicians and health facilities, originating from coast to coast in

North America and overseas.

No one, not even the unusually prescient Dieter Beer could have imagined such intense and early interest, coming from both lay people and health professionals alike, and in the two separate and equally forceful aspects of the young company – the scientific and the commercial.

"I do not understand the business but I do understand the product," Dr. Bounous concedes even now. "The business Chuck is building has begun to distribute the product, which has brought to my desk so many letters and phone calls, questions and reports, from everywhere. I remain busy just trying to answer some of the most interesting medical questions you can imagine.

"One person, for example, wanted to know if Immunocal™ could restore hair growth to reduce male-pattern baldness. I did not have a clue but when I checked the literature on glutathione, I was surprised to find that the hair follicles in this condition, or I should say the basal cells devoid of hair follicles, are associated with low glutathione levels. What could I say?

"Another anxious lady with a problem of infertility, in desperation I guess, wanted to know if there was any hope that she could improve her chances with Immunocal™. You can imagine the big problem of trying to understand the role of immunity or even autoimmunity, in infertility. I wrote back to her that one in three infertile couples find solution just in pursuit of a solution, that is, even without specific intervention. Why? Who knows? Many such remedies may be a placebo effect but ask any of those couples and they would reply 'who cares?'

"Most of the time handling all these queries, I feel I am outside my expertise. I was trained as a surgeon and drifted into nutrition and immunology but I am not an internist or a psychiatrist. The basics of Immunocal™, glutathione and the immune response is really all I know. Fortunately, the benefits to be derived come from nature and not my expertise. But I am having real fun anyway, just dealing with the wide range of inquiries, nevertheless."

One recent phone call nearly made Dr. Bounous drop his professional skepticism and allow him to feel truly happy and excited. The caller was a neurologist from one of America's most prestigious medical schools, who had himself contracted ALS (Lou Gehrig's disease) which had more or less confined him to a wheelchair. He had been introduced to Immunocal™ and had been taking it for some months.

Now the physician was calling to report his own anecdotal story that he had experienced significant improvement. He could actually weight-bear much better and even travel again, and was back at work part-time, too. The neurologist expressed his gratitude for whatever role Immunocal™ had played in his remarkable progress, but more importantly for science, he desired to conduct a clinical trial, using Immunocal™ with certain other ALS sufferers, intending to follow each case himself.

"That was the call in which everything seemed to hit home for me," he said. "I had wanted to develop something helpful and effective, something useful – and this doctor, one of my own peers no less, assured me that my product did have practical benefits even in his own case. I could hardly ask for more."

Perhaps that was the moment in which Dr. Bounous at last figuratively emerged from the world of white coats and solitary research. He would turn his investigations into another reality – that of the large numbers of human beings, sick or well, who should (and probably many will) benefit from Immunocal™. The doctor is ever so slow to believe in his own success. He is a scientist in reluctant transition. The research has been done, or at least his contribution has been made – now it's time for application, for promotion and distribution. This is a totally different challenge.

Gazing out of the doorway of his office into the stream of Immunotec staff members walking past, he seems almost astonished to find himself in such a business setting.

"This is less stressful," he comments, "Research is always stressful because you are waiting for results. You cannot change them.

Research can result in bad news if nature so dictates. But business people deal with human beings. They can effectively change the outcome."

Clearly Dr. Bounous enjoys the 'people part' of his new life. He is making up for lost time. He offered strong input, for example, during the recent visit of a group of Indonesian government officials. Intrigued by medical journal reports of Immunocal's immune-enhancing properties, the group journeyed to Montreal to learn more first hand. The problem: Indonesia's high incidence of infectious diseases and mounting health care expenses. Could Immunocal™ help Indonesians prevent some of these illnesses? Bounous, fascinated by the question and flattered by their travels from across the globe, spent hours consulting with the group. At the end, a plan had been decided upon. In one of Indonesia's many small islands, half the population would receive Immunocal™. After a year, a cost/benefit assessment will be made: Immunocal™ as a possible protection versus drugs as the normal treatment.

This type of pragmatic approach is in response to a very real, present day problem – the exorbitant costs of drugs world wide. It should produce some noteworthy results. In a time when "health care" too often means "illness, diagnosis and on-going treatment", Dr. Bounous sees some small encouraging signs that viewpoints all around the world are changing. As always, he thinks 'it is better to *keep* good health than to *restore* oneself from bad health.'

NUTRICEUTICALS

Another encouraging event occurred in April, 1998 when Dr. Bounous and Chuck Roberts testified before a select committee of the Canadian Parliament about government regulations concerning the marketing of natural food products with perceived health benefits like Immunocal™ . It is a consummate example of the new category of 'Nutriceuticals', as some in the industry now call them.

In Canada there are only two recognized broad categories

for products normally ingested by mouth: foods and drugs. The law is very prohibitive regarding any claims for 'food' products and naturally demands much more justification, specification and regulation in the distribution and use of 'drugs'. In the U.S., a third category has now emerged, that seems to bridge these two poles. 'Nutriceuticals' are considered to be nutritionally **(nutri-)** derived as in food sources, and are nevertheless demonstrated to have some clinical, therapeutic or pharmaceutical **(-ceutical)** like effect when taken on some regular basis.

It is becoming increasingly clear and the public is making growing demand that this new category of product be legally recognized and yes, regulated. In the United States, there already has been a government response and changes continue to be made in the FDA regulations.

Immunocal™ is clearly a leading candidate for this nutriceutical category. It is carefully derived entirely from a basic staple food source – 'cool, refreshing milk' whey – that provides a concentrated source of the three key proteins rich in cysteine dipeptides and maintained in undenatured form. The proven benefits of this novel product demonstrate its clinical effectiveness in immune-enhancement and much more. It therefore satisfies the two essential criteria for this new category definition.

Armed with a stack of publications and good research data, as well as method-of-use patents, Dr. Bounous and Chuck Roberts made a formidable presentation of the case for Immunocal™ before the Parliamentary Committee and they were well received. Nothing is as formidable as good, credible, scientific research and that's on the side of Immunocal™. But politicians are as interested in form and appearance as much as in fact and substance.

"I changed my opinion of politicians," Dr. Bounous jested. "These gentlemen and ladies came well-prepared and asked pertinent questions. Such interest makes me believe we will see changes made and natural products like ours will no longer have to be medical step-children in North America."

As a counter point to such progress, however, the press has knocked at Immunotec's door more than once, seeking to expose what are presumed to be unconfirmed medical claims. Fortunately, once the diligent journalists see the published research reports from reputable medical journals, see the credentials of the investigators, see results of clinical trials and talk with Dr. Bounous and Mr. Roberts, they invariably go away without a sensational story, but with a new appreciation for a serious company which understands that until the laws and regulations are changed, it cannot legally and responsibly make medical claims about any 'food' product.

No medical claims per se. That's what the law says. That's what the company wants. Therefore, that concept is drilled into each Immunocal™ entrepreneur from day one. Anecdotal evidence by now abounds, but Dr. Bounous, the king of the skeptics, insists that only laboratory and clinical test results are justifiable and therefore reliable and useful. He devoted his life and career to research and so he accepts nothing less than the scientific method as authoritative. He is relentless in his pursuit of the truths of nature and he is passionate about the results that he has painstakingly found.

Ironically enough, others with competing interests and less credible support for their claims have not hesitated to use his work and misuse his results to make sweeping generalizations which have no scientific foundations. The consummate example is in the use of the term 'whey protein'. Of course Dr. Bounous used a 'whey protein concentrate' in his laboratory experiments and subsequent early clinical trials. And obviously, the patents issued to Immunotec Research, Ltd for the dramatic method-of-use claims pertain to "whey proteins". However, Dr. Bounous has also demonstrated and emphasized the critical need for the stringent proprietary technology and quality control that is necessary to provide the necessary **undenatured** proteins. Molecular biology and cellular physiology tend to be unforgiving. Cysteine will successfully travel to the target cells only as dipeptides. Cysteine will most effectively enter the cell, across the discriminating cell membrane, only in the dipeptide form. No ordinary separation of these delicate proteins, rich in the cysteine dipeptides, will afford the effective and reliable results we seek.

To illustrate this point consider the experimental results in Fig. 4. It shows the effective immune enhancement in animals fed Immunocal™ for a period of three weeks. Again, that effect is attributed to the increased production of glutathione in the lymphocytes through the dietary provision of supplementary doses of the GSH precursor, cysteine, in the form of the cystine dipeptide. More particularly, this is in contrast to the negative results with commercial whey protein concentrate or casein.

The emphasis always is on *bioactive* whey protein. Immunocal™ is the only one which actively enhances significantly the vital glutathione levels inside the cells of the body. 'Whey proteins' as a general term describes a wide variety of possible compositions that do little justice to the patented claims. **All whey proteins are not demonstrably equal**.

The contrasting physical-chemical characteristics and biological activity of different types of whey protein concentrate is further summarized in Table 5. The effective superiority of Immunocal™ is again evident. But Dr. Bounous is reassured by the conviction that good science will prevail with or without legal defense of his breakthrough product. Only reliable results by competent researchers in the laboratory and in the clinic will verify the truth of each claim in time. Such results have begun to accumulate.

ON - GOING STUDIES

A McGill University *in vivo* study of prostate cancer patients is monitoring PSA values to examine the effect of Immunocal™. The trial is proceeding well and the study has been extended to even advanced, intractable cases.

A clinical trial on the effect of Immunocal™ in patients with cancer of the breast is planned at some major treatment centers in the United States.

FIG. 4

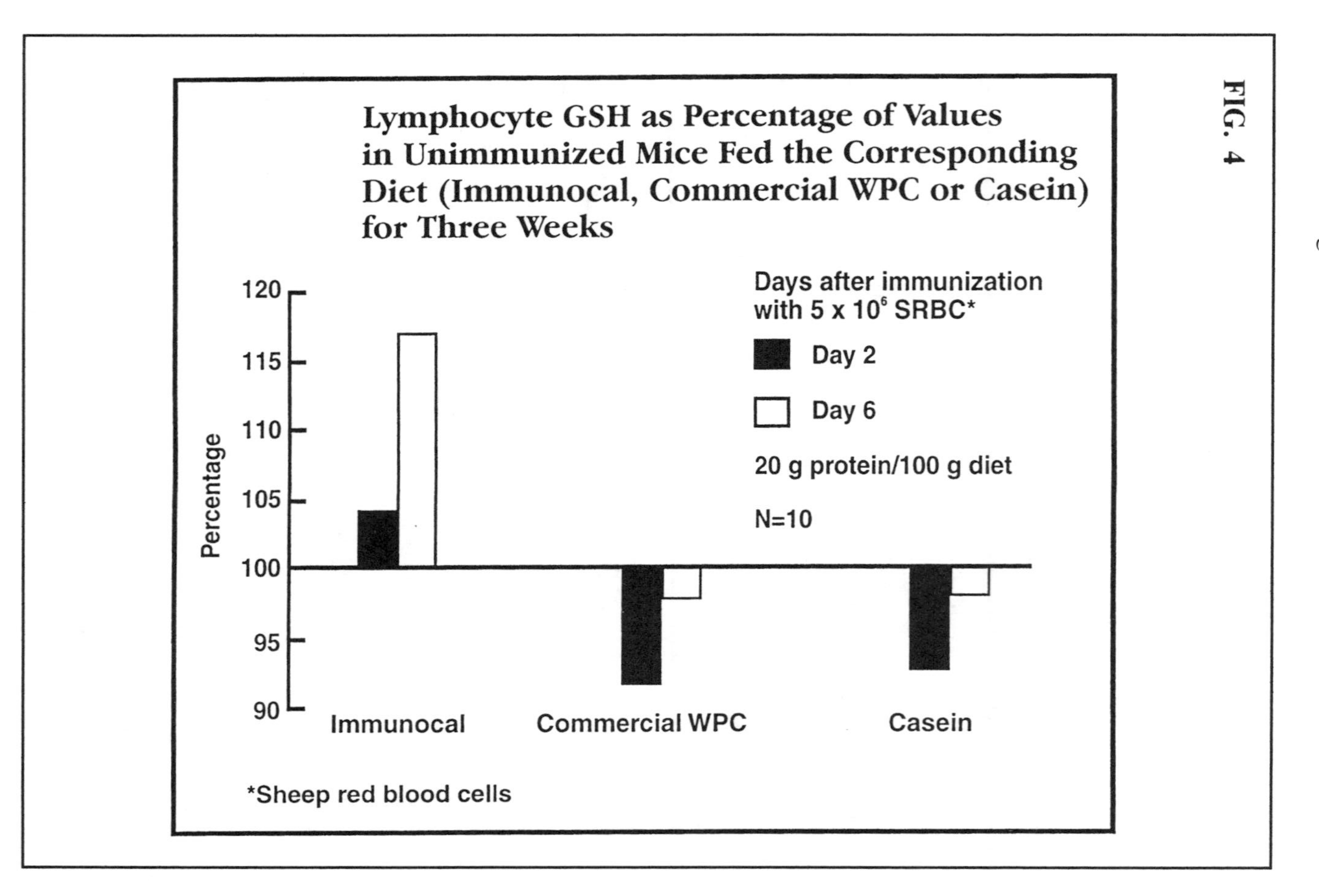
Lymphocyte GSH as Percentage of Values in Unimmunized Mice Fed the Corresponding Diet (Immunocal, Commercial WPC or Casein) for Three Weeks
Days after immunization with 5 x 10^6 SRBC*
Day 2
Day 6
20 g protein/100 g diet
N=10
Percentage
120
115
110
105
100
95
90
Immunocal
Commercial WPC
Casein
*Sheep red blood cells

TABLE 5.

Physical-chemical characteristics and biological activity of different types of whey protein concentrate

	Undenatured conformation		Effect of 3 weeks dietary treatment: Glutathione (µmol/g)	
	Solubility index (pH 4.6)	PFC [1] X 10^{-3}	Liver	Heart
Immunocal	99.5%	148 ± 16	7.95 ± 0.40	1.15 ± 0.7
Product 1	97%	65 ± 14	6.64 ± 0.41	1.0 ± 0.7
Product 2	97.1%	66 ± 17	6.04 ± 0.36	-
Product 3	96%	44 ± 15	6.70 ± 0.20	1.01 ± 0.5
Product 4	95%	67 ± 16	-	-
Product 5	98%	31 ± 8	-	-
Product 6	90.1%	65 ± 20	-	-
Casein	-	35 ± 9	-	1.0 ± 0.8

Values are expressed as mean ± SD;
[1] Number of plaque-forming cells/spleen (antibodies) 5 days following immunization with 5 x 10^6 sheep red blood cells, 20 g protein/100 g diet fed to mice during three weeks before and following immunization.

This table illustrates the immunoenhancing (antibody production) and glutathione promoting activity of different types of whey protein concentrate (WPC). The above data compare the activity of Immunocal to other commercially available WPC.

When we look at all the factors described in textbooks in relation to cancer, such as tumor necrosis factor (TNF), interleukin (IL-1-18), oncogenes, etc, we are reminded of the work by epidemiologists from Harvard, who demonstrated a decade ago that there has been little significant time change achieved between diagnosis and death for the major types of tumor. They were provocative enough to suggest that much cancer research has been following perhaps the wrong track and should probably be abandoned. This sad concept was reiterated recently by the Chief of Oncology at the University of Illinois in a TV interview.

New approaches are certainly warranted. But they must still be characterized by good science. Our results on cancer to date do cause us to wonder if the special whey protein concentrate Immunocal™ does not interfere with an intimate and crucial part of carcinogenesis or cancer proliferation which is still totally unknown to most current oncologists.

A large scale Phase III Canadian AIDS Network study is in progress with Dr. R. Lalonde as principal investigator. In addition, a recent proposal has been made by a dozen leading AIDS specialists in New York, that a combined study protocol be devised for a national clinical trial, to examine the role of Immunocal™ as an adjuvant in the present standard triple-drug cocktail therapy for HIV infection. It has been suggested from a couple case reports (observed by Dr. Bounous too) that the concomitant use of Immunocal™ sustains the glutathione levels which otherwise become depleted by the powerful drug cocktail. The net effect is better tolerance and patient compliance.

Dr. Larry Lands, Assistant Director of Respiratory Medicine at McGill has been studying exercise responses in healthy young adults and those with cystic fibrosis. The study is geared to investigate the possible role of Immunocal™ as a sports supplement. It has carried over to the cystic fibrosis (CF) phase with such positive results that Dr. Lands wrote – "There is tremendous excitement in the CF community to the point that I have been asked to write about Immunocal™

... In addition, there is an increasing number of articles appearing in the scientific press on the use of antioxidants in slowing the progression of or actually reversing lung disease, such as CF, chronic bronchitis, emphysema ... we see an expanded role for Immunocal™ in these conditions ... furthermore, Immunocal™ is a well-tolerated nutritional supplement that has the appeal of being a natural therapy, rather than being perceived as a medicine."

Just to illustrate the import of Dr. Land's observations, an isolated case study is summarized below:

A 41 year old woman had been a patient at the Royal Victoria Hospital, McGill University, Montreal since she developed Hodgkin's Lymphoma. This required both chemotherapy and radiotherapy to her chest. She subsequently developed pulmonary fibrosis and respiratory insufficiency as a result of her treatment. She was followed by the hospital's pulmonary service for this deteriorating condition by a senior staff member of the department.

Her pulmonary function tests showed a progressive and significant decline from Feb/95 to Nov/97 which left her quite short of breath and functionally distressed. She unfortunately did not respond to early treatment modalities. She commenced taking Immunocal™, 20 grams/day, and in six weeks she no longer suffered from her debilitating dyspnea. Objective pulmonary function testing March/98 demonstrated a dramatic return almost back to her pre-morbid state.

Another Phase II Double Blind study of post-operative immune status is taking place in Munich, Germany, in which Immunocal™ is being compared with another whey protein concentrate as control.

One interesting on-going study that must also be mentioned is that of internist Dr. Paul Cheney at the Cheney Clinic in North Carolina on Chronic Fatigue Syndrome (CFS). He has a worldwide catalog of thousands of CFS patients' files associated with his clinic. He already knew of the reduced intracellular glutathione in CFS patients and had initially tried giving glutathione supplements to a few patients, in the absence of anything else. When he learned of Immunocal™ he inquired about doing a small study which is now nearing completion.

The bottom line of all this research activity, is that Immunotec Research, Ltd. will remain true to its name. It is a research organization, first and foremost.

"We must not stop doing research," Dieter Beer cautions. "That is the only way to maintain our jealous credibility and to keep abreast of health developments. We must be on the cutting edge. We must set the pace for immune system boosters and not follow any other leader in the marketplace."

Immunocal™ was highlighted at the World Conference on AIDS in Geneva, Switzerland in the summer of 1998. Meanwhile, a dozen American physicians, operating as a team but from various locations from New York to California, are designing the national clinical trial in which Immunocal™ will be administered as an adjunct to traditional AIDS medications.

NORMAL ADULTS

Prevention must be the important concern for all normal healthy adults. Even in the absence of clinical signs of illness or disease, glutathione is still at work. But the question arises, ***can supplementation with glutathione precursors lead to up-regulation or increase in tissue glutathione levels for the average healthy person?***

In a recent double-blind clinical study, the effect of supplementation with Immunocal™ on glutathione (GSH) levels and muscular performance in normal adults was observed by Dr. Larry Lands and co-workers (73). Since oxidative stress contributes to muscular fatigue and GSH is the major intracellular anti-oxidant, they hypothesized that supplementation to modulate GSH should enhance performance.

Twenty healthy young adults (10 males) were studied pre- and 3 months post-supplementation with either Immunocal™ (10 gm twice per day) or casein placebo. There were no initial baseline differences between the two groups (i.e. in age, height, weight, % ideal weight, Peak Power or 30-sec Work Capacity). One placebo subject dropped out, and one Immunocal subject's follow-up results were technically unacceptable. Lymphocyte GSH was used as a marker of tissue GSH. Muscular performance was assessed by whole leg isokinetic cycle testing, measuring Peak Power and 30-sec Work Capacity.

What were the findings?

(i.) Lymphocyte GSH increased significantly in the Immunocal™ group (37.8±12.47%, $p<0.03$) with no change in the placebo group (-0.9±9.6%).

(ii.) Both Peak Power (mean rise 13±3.5%, $p<0.02$) and 30-sec Work Capacity (13±3.7%, $p<0.03$) increased significantly in the Immunocal™ group, with no significant change in the placebo group.

These results have clear implications.

Since normal healthy adults show such marked increases in glutathione levels when supplementary Immunocal™ is used, this self-regulated protective system (with its own feedback inhibition) may be operating "below its optimum" level, not only in persons with disease, but also in normal, healthy individuals.

ATHLETIC PERFORMANCE

The significant increase in physical performance observed also supports the use of such a well-tolerated whey-based cysteine donor to augment muscular performance. Other indices of well-being and fitness which are influenced by ongoing oxidative stress, may also be benefited by the same glutathione modulation.

As the major clinical indications for Immunocal™ continue to be investigated in the various types of studies just mentioned, there is therefore a need to remain focused on the important role of glutathione and hence Immunocal™, in primary disease prevention and wellness. After all, the key function of the immune system is to be pro-active. It seeks to be constantly on the alert to detect the enemy and mount a swift and effective defense in the pre-morbid condition. It is still worth repetition that 'an ounce of prevention is worth more than a pound of cure.'

That's at least one more reason why Immunotec is blessed to have a keen athlete and fitness-buff as a principal on staff. John Molson is particularly interested in Immunocal's wellness-enhancing effects. As the above results show, those can be important and dramatic. **These efforts in normal healthy adults underscore the universal market for Immunocal™.**

It is amazing how attuned serious athletes are to their bodies' performance. They know their capabilities and their limits. They know what each activity costs in physical abuse. They bear the burden themselves and learn to cope with the consequences. Reducing oxidative stress decreases the pain factor after strenuous activity and shortens recovery time.

"The key thing for athletes is training time and fitness levels," according to John Molson. "If you can reduce sickness and down time, you increase an athlete's physical best. We have seen elite athletes with no colds, or flu, and we know that glutathione actually 'medicates', thereby decreasing muscle soreness.

EXPANSION

Immunocal™ product research continues. Dr. Bounous has long since removed all fat and lactose from the whey product, making it appropriate for even the lactose-intolerant individual, as well as, and perhaps more importantly, free of fat - and water-soluble pesticides and other chemicals. He continues to investigate the medical uses of whey, projecting such possibilities as its use with gum surgery, for example.

As Bounous and Molson have been focused on the product in the early start-up phase of Immunotec Research, Ltd., Dieter Beer and Chuck Roberts have been very active in developing markets for it. Together there has been balance and synergy to the whole operation.

Licensing arrangements are rapidly being made on all continents. Each country has its own regulations but the need for Immunocal™ remains a common factor. Its benefits should be experienced by people living in Australia, Brazil, China, Denmark, England, France, Germany, Hong Kong, India, Japan ... and eventually around the globe ... to what was until recently known as Yugoslavia, and even to Zimbabwe and Zambia. With Dieter Beer and Chuck Roberts at the helm nothing short will do.

In 1996 Dieter Beer negotiated an agreement between Immunotec Research, Ltd and one of Germany's largest and most aggressive pharmaceutical companies, Fresenius AG, to become Immunotec's exclusive licensee for distribution of Immunocal™ throughout Europe and Africa.

This represents a major coup for the company, of course, since Immunocal™ will be positioned in Europe as a significant medical breakthrough in cell defense. One can only hope that Canada and the United States will soon broaden their perspectives in such matters. Indeed, the U.S. National Institute of Health in 1993 formed an Alternative Medicine branch which promises to help do just that. As the research data accumulates, the case for Immunocal™ as a safe, effective and convenient enhancement to the immune system – that and much, much more – should become overwhelming and soon irresistible (at least, to the public mind).

For an immunologist like Dr. Patricia Kongshavn, the issues

are simple.

"Immunocal™ is a detoxifying agent and an immune system booster. It is good for the lungs, skin, eyes – particularly the surface of the eyes – the kidneys and liver — and more.

"I am watching two children who are being medicated for lymphoma. With Immunocal™ they are experiencing no sickness and no hair loss from the strong drugs.

"When I look at the whole picture I am fascinated by it. Immunocal™ is unlike anything else. I know it works."

She should know. She was there almost from the very beginning.

BACK TO THE BEGINNING

This story actually began in a small, drab medical-office-turned-coffee-lounge at the Montreal General Hospital, a teaching hospital and research facility affiliated to McGill University in Montreal, Quebec. Here, among the shelves of medical books and memorabilia which belonged to Dr. Bounous' early friend, boss and mentor, Dr. Fraser Newman Gurd, Dr. Bounous '*accidentally*' found the scientific key which opened the door to intracellular glutathione modulation.

Today the place looks much the same. An ancient iron bookstand holds a worn 1933 edition of the Shorter Oxford English Dictionary, *exlibris* Fraser Newman Gurd. Here are Dr. Gurd's stethoscope and medical bag, an old syringe, prescription pad, and a volume entitled "A Century of Surgery 1880-1980."

Only a few miles away from downtown Montreal his protégé, Dr. Gustavo Bounous, lives out an exciting and very fulfulling new chapter of his own scientific career.

Bounous, the consummate career scientist, continues his investigative work, begun at McGill so many years ago. But now his world has widened. No longer alone, professionally or personally, he enjoys living on the fringes of the business world, while continuing his own absorbing professional interests as part of his work with Immunotec Research, Ltd.

Each of the company principals is a family devotee. Photos of spouses, children and grandchildren adorn desks and bookcases everywhere, and now Dr. Bounous's area is no exception. The once somewhat reclusive scientist, never married, never head of a household, for the past two years has enjoyed taking care of a young ward who brightens his home. Her picture radiates from the small frame which Dr. Bounous displays proudly atop a prominent shelf in the corner of his small office.

Dr. Bounous has known fourteen-year-old Genevieve DeSerres all her life. A close friend of her grandparents, he helped little Genevieve early on with her homework and took an avuncular interest in her little personality and life.

At age twelve, the girl's life changed radically when her parents divorced. Because she did not like the environment of possible boarding schoool, Genevieve independently elected to come stay with Dr. Bounous during school days. The bachelor doctor moved to a larger place closer to the girl's school and took to surrogate fatherhood like a duck to water, praising the "little girl's" every accomplishment.

"I am so privileged today. I think of myself now as the only grandfather who has never been a father." He says that with such pride and pleasure.

"She led me with her little hand into a wider world, a child's world with a million different interests," the doctor says. "She is so smart, and so nice and intelligent."

Just next to his office, Lina Roti, a young lady from southern Italy, and a Chartered Accountant with a degree in Mathematics, has her office. She is Financial Controller of the company and assists in many other ways which help to simplify Bounous's busy life. And in a similar way, from just outside Toronto, Dieter Beer's very competent executive assistant, Claire McKenna, helps Dr. Bounous overcome what he calls his classic absentmindedness. She has been assisting the doctor in a number of practical ways for the past ten years.

"I admire the doctor's modesty," Claire McKenna said. "He is so humble, a man who derives much gratification from helping people. But it's amazing that such a brilliant person should be so absentminded!"

Dr. Bounous says he lives for the moment. Now AIDS is being studied, cancer researched, infections treated. Now medical outlooks are shifting and the community is responding. Now people worldwide are taking more responsibility for their personal health.

Immunocal™, Dr. Bounous believes, will contribute to a significant degree to giving many people the opportunity to give themselves a real boost in health, and a complementary alternative in many cases of sickness and disease. This is practical science ...

" *I did not discover anything.* I was privileged to find a choice protein mixture, carefully derived from milk serum so that it remains undenatured and containing the critical precursor of glutathione," Dr. Bounous describes Immunocal™ with affection and pride.

He has all the right to do so. He led this breakthrough.

To sum up Dr. Bounous's life work, what better than his own words:

"Unlike specific antiretroviral drugs which may induce mutation, hence resistance of the virus to therapy, the normalization of the lymphocytic glutathione levels and redox status through a cysteine delivery system, represents a totally different approach through which the natural cellular defense system is boosted and against which the virus cannot apparently build up resistance by mutation."

Dr. Bounous's hope remains for the ultimate triumph of the immune system. His own breakthrough contribution to cell defense has been made.

20.

YOUR BREAKTHROUGH IN SELF-DEFENSE

This book was written with you the reader in mind. It is clear that *Breakthrough in Cell-Defense* has important implications for anyone interested in health and wellness and may be useful as adjuvant therapy in some conditions of illness and disease.

But this is only the beginning of the story. With more research and the increasing use of Immunocal™, a much better understanding and appreciation will be gained from this new unveiled secret of nature. But enough is known already about this vital health link and the correlation with humanized milk serum to justify its widespread use. There is no question of the paramount value of traditional breast feeding for the healthy immune system of babies. It is as if nature designed the best means of protection for the most vulnerable of the species. There is no adequate substitute for mothers' milk. This *breakthrough* in effect, now makes some of its key benefits available to the entire population.

The studies to date of clinical applications of the *breakthrough* have shown positive results in a few select patients suffering from a wide variety of illnesses: AIDS, cancer, viral illnesses, chronic fatigue, respiratory diseases, just to name a few. Surely, a lot more work remains to be done and further research is already in progress. But it is already apparent that modulation of intracellular glutathione offers a new and effective adjunct to patient management. The clinicians involved in patient trials are showing increasing interest in this novel approach to cell-defense. All the results to date auger well for the future.

However, the real value of Immunocal™ is not in its pharmocological action, or the detailed biochemistry of its function, and certainly not in the specifics of the proprietary technology used to retain the specialized proteins in their undenatured state for consistent bioactivity. This is reserved for the scientists.

Of course, all of that is very important and research must continue. Many more questions still remain. But the usefulness to each reader is in the individual person choosing to add this safe, proven breakthrough product to their daily dietary regimen as a new lifestyle habit. There is an immediate benefit to be gained since improved **cell-defense is the best form of self-defense.** You can therefore take advantage of this even now, and so help to optimize your own immune system.

YOUR IMMUNE SYSTEM AT RISK

It is clear that the key to prevention and treatment of diseases and even common illnesses is a healthy, well-functioning immune system. We were born with this wonderful defensive apparatus. Many were provided from birth with the essential glutathione precursors present in their mothers' milk. But with maturing age, changing lifestyle habits and an increasingly antagonistic environment, this source of protection may not be as guaranteed as it once was.

Think for just a minute of the hundreds of new chemicals introduced into the atmosphere, the water supply, and into agriculture and foods. Or, think of the emergence of new strains of lethal bacteria and viruses, with the widespread use of antibiotics in the twentieth century. Or, think of the pace of life and the ironic sedentary lifestyle of contemporary society. Add to this, changing dietary habits, a modern sexual revolution, the popularity and abuse of drugs (prescription and non-prescription) and the picture gets no better.

Your own immune system could be at risk. You may be aware of this after burning the candle at both ends with limited rest or sleep. You may be suspicious after specific exposure or a deprivation diet. You may have succumbed again surprisingly to an acute infection, or even frequently do so. You may have experienced trauma, or surgery or chronic illness. You may be an exhausted athlete. You may be suffering from chronic degenerative change or fatigue. You may have any of a hundred and one other reasons to be concerned. In any case, your immune system can sustain damage and become defective. Consequently, it will not do its job effectively.

Free radicals play a role in most common illnesses and many degenerative diseases. More and more, different oxidation mechanisms are being implicated in disease processes. Hence, the so-called ***antioxidant revolution*** is already taking place, with glutathione becoming the justifiable leader of the pack. When these versatile protectors are depleted, the body's defense is then too weak, and it becomes overwhelmed. Your performance suffers.

Scientists today believe that all of us (and that includes you), on a daily basis are victims of microscopic damage to our cells and even our organs. Random mutations (mistakes, changes) take place during your normal growth and reproduction. Renegade cells must be contained and destroyed. Your healthy metabolism produces toxic products that must be neutralized. Your cellular damage from free radicals, oxidation reactions and ingested or inhaled poisons must all be counteracted regularly. Only a well-functioning immune system can defend your body from all such insults to preserve your health. Otherwise, you are prone to illness and disease. The recent AIDS epidemic is only the most extreme demonstration of this. In the real world, depressed immunity is pandemic.

The immune system can become harmed in a short period of time, especially if we abuse ourselves. It is genetically designed for your protection – to shelter you from disease. But it will protect you best, only if you do your part. This relates to all the lifestyle issues. You hold the key. You choose your habits.

Your immune system actually has different responsibilities. One is obviously to fight-off any internal source of illness. Another is to gather information on any developing disease process and to communicate rapidly to different but specific cells and organs which can then be recruited for defensive function. Other cells produce 'ammunition' (antibodies, cell toxins, etc.) which destroy foreign invaders before they strike. The intricacy of the whole system keeps you safe and healthy, despite the threats of infection from bacteria, viruses, yeast, fungi and even protozoa. Renegade cells are neutralized early. Other toxins are kept at bay, and tissue damage is averted. The immune system, with its specific cells scattered throughout the body, has the unique power to recognize chemical and biological threats to survival. These threats are then devoured by so-called phagocytes. They are systematically isolated first, then neutralized and finally destroyed.

How kind are you to your immune system? Do you make it work twice as hard just to cope with your particular lifestyle habits. Are you cognizant of the needs of your immune system? Are you aware of the risks that can be posed? Are you informed about what you can do to boost your own immune system? Or, do you just take it all for granted?

Low immunity can manifest itself in all kinds of sub-clinical symptoms: including the common cold, flu-like symptoms, low-energy, depression, fatigue, mood swings, common susceptibilities, fevers, mild gastrointestinal upset, etc. In fact, recurrent illness, or just consistent common complaints, or maybe less than peak performance may be obvious indications. Common excuses for 'degeneration with age' often reflect a deteriorating immune system. Research has shown that even serious or professional athletes who are exposed to oxidative stress can experience increased susceptibility to illness subsequent to work-out/exercise.

What about you? Are you interested in better health?

So what can you do?

OPTIMIZING YOUR IMMUNE SYSTEM

You can strengthen your own immune system by balanced and adequate dietary intake, probably including a multi-vitamin mineral supplement. That's a great place to start. Then include some moderate exercise; adequate rest; controlled stress; avoiding alcohol, drugs and smoking; keeping positive attitudes; minimizing exposure to physical, chemical and biological insults, and having a periodic health examination with your doctor. Just reflect on the following :

A 10 - POINT PRESCRIPTION TO OPTIMIZE YOUR IMMUNE SYSTEM

1. **Diet.** There is no substitute anywhere for a healthy diet that is varied; high in fiber and low in fat; rich in fresh fruits, vegetables and whole grains; abundant in fluids and restricted in sugar and salt. Modern fast and convenience foods may not be doing any justice to the immune system.

2. **Supplementation.** To this cornerstone of good nutrition, it makes sense in contemporary industrialized societies to add supplements to ensure a guaranteed supply of some essential nutrients. In particular, Vitamins A, C and E have special importance to the immune system, as do the trace minerals selenium and zinc. There is also contribution from B Vitamins. Glutathione enhances or regulates the antioxidant vitamins to make them most effective.

3. **Exercise.** Nutrition is a great place to start. Then include some moderate exercise. Muscle activity is a process of aerobic respiration and cellular oxidation. Both mechanisms produce energy by catabolism, but they also produce harmful metabolites and those same dangerous free radicals. Glutathione is consumed during and after such oxidative stress, to preserve the cellular integrity and restore homeostatis. To replenish glutathione improves athletic endur-

ance, reduces painful hangover and shortens recovery time.

4. **Rest.** Adequate rest is indispensable for the body to recuperate from all the physical, mental and biochemical stress that it must endure each day. The rhythm of life was designed for restoration of health on a regular basis. We turn on and start to wind down each morning and then we must turn off and wind back up at the end of the daily grind. Relaxation and re-creation are even as important as sleep.

5. **Controlled stress.** Speaking of stress, there is an ordinate level of tension and excitement that stimulates adrenalin and productivity. But those same hormones (in excess) can cause havoc and activity can become the enemy. Every person has a physical, mental and emotional elastic limit. Beyond that, we snap one way or the other. Learning balance, discipline and resilience will do wonders for your immune system.

6. **Avoiding addiction**. Alcohol, drugs and smoking are notoriously bad for the immune system. Enough said.

7. **Attitudes.** Evidence is accumulating that the way we think and react to our life situation is somehow coupled into our physical bodies. This area of neuro-psycho-immunology is in its infancy but it is clear that remaining positive, hopeful, caring and interactive has definitive consequences for health and disease. You choose so much of your own destiny.

8. **Exposure.** You will want to minimize exposure to all the physical, chemical and biological insults in the environment - as much as you can. These can all wreak havoc on your immune system. Do a brief analysis of what may be threatening to your health in your home and work surroundings and take appropriate measures to remove them.

9. MD Visits. There is no substitute for the periodic health examination by your local doctor. Many a serious condition can be avoided by routine screening, assessment and pro-active consultation. This is a major contribution to optimizing your immune system.

10. Immunocal™.

And now... thanks to this major discovery, you can enjoy the benefits of the new **Breakthrough in Cell-Defense** - and take part in **the *real* Glutathione Revolution** – by taking Immunocal™ on a daily basis. You will have nothing to lose and so much to gain. It will help to optimize your all-important cellular glutathione level and boost your own immune system. Again, 'an ounce of prevention is better than a pound of cure.'

Remember, inside the cells, glutathione is the key protective molecule of your immune system ... *where* you need it most. For most effective immune response, Immunocal™ can provide a key source of cysteine for glutathione synthesis ... *when* you need it most. Glutathione synthesized inside the cell is nature's best 'mop' to clean up the mess. It is active in health and disease for your protection.

The typical person can now take advantage of this safe, convenient, effective and economical method to enhance the glutathione production by their cells whenever necessary. This may be just for the maintenance of health or for possible adjuvant therapeutic value in many disease states.

Here then is *the* key that unlocks the door to *all* the known benefits of glutathione inside the cell. Immunocal™ is registered and patented as a food supplement in North America. No prescriptions are therefore necessary, but it may be used therapeutically as prescribed by a physician.

Anyone will benefit from the possible optimization of glutathione levels and the clinical correlation of that result with the immune system and more. There truly is no substitute for strengthening the body's defenses and general resistance to all the daily stresses and threats from without and within.

You can lead a healthier life. You can now optimize the glutathione inside your cells. That is nature's way of cell-defense and that is your best option in self-defense. Indeed, it bears reiteration, ***your best self-defense is cell-defense.***

The body of research already published on the role of glutathione in health and disease, when coupled with this new breakthrough in simply and effectively enhancing its production *in vivo*, will have major consequences for public health and in the medical management of patients for the foreseeable future. It is a late-breaking but promising approach to exploit nutritional pharmacology as another alternative for both primary and secondary prevention strategy.

On-going research will continue to broaden its application into the twenty-first century.

EPILOGUE

When I began this book project, I was aware that we were setting out to document for a wider general audience, the true story behind a man and his mission. That man had dared to pursue some unpopular research paths where his own curiosity and experimental results led him. He was committed to the scientific method and labored diligently, often in isolation but also in collaboration with other scientists and later with businessmen too, until his surprising discovery would come of age and find practical significance.

But what I did not anticipate was the distinct privilege I would have to get to know a truly great scientist, in the noble tradition of curiosity, dedication, integrity and modesty. I learnt by precept and example from this research veteran, not to take oneself too seriously, but to enjoy whatever one finds to do and follow wherever the truth leads.

The truth did lead Dr. Bounous to a new mission. Having demonstrated the new 'safe, effective and convenient' way to modulate intracellular glutathione, he became fascinated by the clinical implications and later by the commercial opportunity. But he is still focused on good science, responsible protocols, appropriate analysis etc. He wants everything done right, even if not right now.

As such, the doctor believes in the inevitable triumph of truth and is committed to that. But he senses the world in need and waiting for a universal immune boost that can be made available for widespread use by all categories of people. He believes that what his early 'elemental diet' could do (and now does) for very ill patients, his unique *undenatured* whey protein concentrate can do (and will do) for the general population, and with relevance to *both* health and disease. The mission then becomes a challenge to get the message out, across the professional mainstream and beyond, and to market this new consummate *nutriceutical* to the many who could stand to benefit from this revealed secret of nature.

Just imagine that after all these years, we now have available a simple, commercial, natural product that can deliver such essential benefits of mother's breast milk to the entire population. It is a humanized milk serum indeed. We now understand more clearly how the importance of cell-defense is underscored and exploited in nature. The essential glutathione precursors are adequately provided for the fragile infants. That's what they need for self-defense and that's what they get.

I see clearly now the value of the critical undenatured whey protein concentrate. Its enormous potential in both primary and secondary prevention does open a floodgate for the so-called 'wellness movement.' It sets an entirely new standard for the food supplement industry. That standard now demands good science, peer-review, documentation of clinical claims, stringent quality control and responsible promotion. Nothing less will do.

It never ceases to amaze me how almost everything that is needed for the human solution is provided in nature. Most of the time it is obscured by our own prejudice and stubborn folly. But if we seek, we will find. That's what Dr. Bounous did. He kept on seeking where truth beckoned, and through dogged persistence, he found a surprising breakthrough from a most common source. **As such, he made a true *discovery* - not an invention - for that is truly nature's way.**

However, we must be careful. Too often has 'nature' become a cheap alibi for poor science and inadequate investigation. The uninformed public has often been deceived, manipulated and abused by those who did not do their homework. To be responsible demands that we be rigorous and thorough as we seek to harness the best of nature for the benefit of mankind.

Nutrition is a science of the future. It is a science from nature. But it will take its proper place in health-care only as a science, subject to the same scrutiny and rigor as all other fields of human inquiry and intervention. The power of any such ancillary approach to health and even disease, is found in the awesome wonder of natural life originating in each singular cell. 'Give 'em the tools and they will do the job.' We are all mere simpletons in the face of that inherent sophistication. The reality that Dr. Bounous could discover a way to influence the intracellular response in a such a fundamental way, by modulating glutathione levels, is an elevated beacon of things to come. But further breakthroughs will only be made by similar dedication to fundamental research.

A new concept has emerged with this breakthrough discovery. The idea of nutriceutical modulation of glutathione now opens the door to further explorations of *nutritional pharmacology*. This by no means excludes or negates other more conventional therapies, approved or experimental, for any of the illnesses mentioned in this book. After all, a *breakthrough* by definition, is a 'decisive advance, or discovery, especially in scientific research, *opening the way to further developments.*' We should expect that both basic and clinical research will continue to demonstrate the tremendous potential for the type of complementary management inherent in this innovative approach.

The facile criticism may be raised that the number of cases reported here is so limited that they have no proven clinical significance. After all, contemporary protocols for 'drug' screening and introduction usually demand long, tedious and expensive double-blind randomized clinical trials. And that would be ideal for sure, but it is only a *necessary* pre-condition here if *nutritional pharmacology* had the same attendant *risks* as any other typical 'drug'.

The preliminary and admittedly incomplete data presented in this book are the result of an uphill struggle for clinical research. Most orthodox medical statisticians have typically viewed a *nutriceutical* as a confounding variable in an otherwise strict clinical protocol.

Accordingly, it is only given to these few patients who either reject or cannot tolerate standard therapy. It is then viewed as another 'drug', aimed for example, to kill either virus-infected or cancer cells, rather than as a booster of the cells' natural defense.

The nutritional modulation of cellular glutathione does not impose or even manipulate any *new* biochemical process or mechanism. **It simply delivers what nature demands and facilitates what nature itself controls.** It is therefore fundamentally safe, but it is also effective in supporting the immune reponse.

We are now pleased that the case reports, together with the substantial experimental evidences, were sufficient to convince groups of scientists in different countries to expand this breakthrough research. The Canadian HIV Network is funding a Phase III Trial in AIDS patients who are now receiving Immunocal® as a complement to current anti-retroviral therapy. Similarly, it is being investigated presently as an adjuvent to normal cancer chemotherapy at the Cancer Treatment Centers of America. In Europe, the muli-national pharmaceutical company Fresenius AG, has also committed to large scale clinical trials.

Since the benefits are clear and the approach is fundamentally safe and sound, there is no reason to defer the application of this innovative approach to immune-enhancement. The door that Dr. Bounous has opened will allow others to develop further clinical applications as the widespread benefits become more obvious in the population.

It is no wonder then that when the Medical Research Council of Canada chose very recently to honor the "health science related achievements of a selection of the most talented researchers in Canada" for publication in a documentary book and on the worldwide web, Dr. Gustavo Bounous was among the distinguished few. He has earned his place not only in Canadian science and medicine but on the worldwide stage.

I have been most privileged to get to know him intimately and to co-author this book with such an outstanding scientist of our time.

Yet I sense Dr. Bounous's true greatest lies in his mind and in his spirit. I'm inspired by his modest and unassuming temperament that camouflages such intense dedication to truth and scientific research. He is fascinated by nature and thrilled by this *breakthrough in cell-defense*, but he remains unaffected by its promising commercial success. Ironically, he is most proud of being able to fulfill in some measure, his dear mother's dream.

The road ahead will require much further study, effort and problem-solving. But the proper foundation has now been laid with an emphasis on bold ideas, good science and an unwavering determination to follow wherever the truth leads. I know that Immunotec Research, Ltd. is committed to that goal.

I hope you have captured the spirit of the man behind the breakthrough as you have read this book. May you be inspired by his example, as I have been, to greater efforts, stronger hopes, broader vision and a more abundantly healthy life.

Allan C. Somersall, Ph.D., M.D
Atlanta and Toronto

REFERENCES

1. Jontz J.G., Su C.S., Shumacker H.B., Bounous G., Influence of renal denervation upon renal blood flow. *Surg Gynec Obst* 110: 622, 1960.

2. Bounous G., Shumacker H.B., King H., Studies in renal blood flow: some general considerations. *Ann Surg* 151: 47, 1960.

3. Bounous G., Onnis M., Shumacker H.B., The abolition of renal autoregulation by renal decapsulation. *Surg Gynec Obst* 111: 682, 1960.

4. Bounous G., Shumacker H.B., Further studies on renal decapsulation. *Surg Gynec Obst* 113: 567, 1961.

5. Bounous G., Shumacker H.B., Experimental unilateral renal artery stenosis. *Surg Gynec Obst* 114: 415, 1962.

6. Bounous G., Shumacker H.B., Further study in experimental unilateral renal artery stenosis. *Surgery* 52: 458, 1962.

7. Bounous G, Hampson L.G., Gurd F.N., Cellular nucleotides in hemorrhagic shock. Relationship of intestinal metabolic changes to hemorrhagic enteritis and the barrier function of intestinal mucosa. *Ann Surg* 160: 650, 1964.

8. Bounous, G. Brown R.A., Mulder D.S., Hampson L.G., Gurd, F.N.: Abolition of tryptic enteritis in the shocked dog. Creation of an experimental model for study of human shock and its sequelae. *Archives of Surgery*, 91: 371, 1965.

9. Bounous G., McArdle A.H., Hampson L.G., Gurd F.N.: The cessation of intestinal mucus production as a pathogenetic factor in irreversible shock. *Surgical Forum*, 6: 12, 1965.

10. Bounous G.: Metabolic changes in the intestinal mucosa during hemorrhagic shock. *Canadian Journal of Surgery*, 8: 332, 1965.

11. Bounous G., Cronin, R.F.P., Gurd F.N.: Dietary prevention of experimental shock lesions. *Archives of Surgery*, 94: 46, 1967.

12. Bounous G., Sutherland N.G., McArdle A.H., Gurd F.N.: The prophylactic use of an elemental diet in experimental hemorrhagic shock and intestinal ischemia. *Annals of Surgery*, 166: 321, 1967.

13. Adams S., Dellinger E.P., Wertz M.S. et al: Enteral versus parenteral nutritional support following laparotomy for trauma: a randomized prospective trial. Reduction in complication rate and overall cost of therapy compared with parenteral nutrition. *J. Trauma* 26: 883-90, 1986

14. Moore F.A., Moore E.E., Jones T.N. et al: TEN versus TPN following major abdominal trauma: reduced septic morbidity. *J. Trauma* 29; 916-22, 1989.

15. Bounous G., Stevenson M.M., Kongshavn P.A.L.: Influence of dietary lactalbumin on the immune system of mice and resistance to Salmonellosis. *Journal of Infectious Diseases,* 144: 281, 1981.

15(a) Cunning A.J. & Szenberg A.: Further developments in the plaque technique for detecting single antibody-forming cells, *Immunology*, 14, 599, 1968.

15(b) Lapp W.S., Mendes M., Kirchner H. & Gemsa D., Prostaglabin synthesis by lymphoid tissue of mice experiencing a graft-versus-host reaction. Relationship to immunosuppression. *Cell. Immunol.* 50, 271-281 (1980)

16. Bounous, G., Kongshavn, P.A.L.: Influence of dietary proteins on the immune system of mice. *Journal of Nutrition*, 112: 1747-1755, 1982.

17. Bounous, G., Letourneau, L., Kongshavn, P.A.L.: Influence of dietary protein types on the immune system of mice. *Journal of Nutrition*, 113, 1415-1421, 1983

18. Bounous, G., Kongshavn, P.A.L.: The imunoenhancing property of dietary whey protein concentrate. *Clinical and Investigative Medicine,* Vol. 11, 271-278, 1988.

19. Bounous, G., Kongshavn, P.A.L.: Influence of protein type in nutritionally adequate diets on the development of immunity. In "Absorption and Utilization of Amino Acids". Friedman, M., (ed.), C.R.C. Press, Boca Raton, Florid, U.S.A., Volume 2, pp. 219-233, 1989.

20. Bounous, G., Shenouda, N., Kongshavn, P.A.L., Osmond, D.G.: Mechanism of altered B-cell response induced by changes in dietary protein type. *Journal of Nutrition*, 115: 1409-1417, 1985.

21. Wang C.W., Watson, D.C., Immunomodulatory effects of dietary whey proteins in mice. *J. Dairy Res*. 62: 359-68, 1995.

22. Bounous, G., Kongshavn, P.: The effect of dietary amino acid on the growth of tumors. *Experientia*, 37: 271-272, 1981.

23. Bounous, G., Sadarangani, C., Pang, K.C., Kongshavn, P.: Effect of dietary amino acid on tumor growth and cell mediated immunity. *Clinical Investigative Medicine*, 4: 109-115, 1981.

24. Visek, W.J.: Dietary protein and experimental carcinogenesis. *Adv. Exp. Biol.* 206: 163-186, 1986.

25. Euker, W.E., Jacobitz, J.L.: Experimental carcinogenesis of the colon induced by 1, 2 dimethylhydrazine – dl HCE: Value as a model of human disease. *J. Surg. Res.* 21: 291, 1976.

26. Bounous, G., Papenburg, R., Kongshavn, P.A.L., Gold, P., Fleiszer, D.: Dietary whey protein inhibits the development of dimethylhydrazine induced malignancy. *Clinical and Investigative Medicine*, Vol. 11, 213-217, 1988.

27. McIntosh GH, Regester GQ, Le Leu RK, Royle PJ. Dairy proteins protect against dimethylhydrazine-induced intestinal cancers in rats. *J Nutr* 125: 809-16, 1995.

28. Bounous, G., Batist, G., Gold, P.: Whey proteins in cancer prevention. *Cancer letters* 57: 91-94, 1991.

29. Birt DF, Baker PY, Hruza DS. Nutritional evaluations of three dietary levels of lactalbumin throughout the lifespan of two generations of Syrian hamsters. *J Nutr* 112: 2151-60, 1982.

30. Birt DF, Schuldt GH, Salmasi S. Survival of hamsters fed graded levels of two protein sources. *Lab Anim Sci* 32: 363-6, 1982.

31. Bounous, G., Gervais, F., Amer, V., Batist, G., Gold, P.: The influence of dietary whey protein on tissue glutathione and the diseases of aging. *Clinical and Investigative Medicine*, 12: 343-349, 1989.

32. Bounous, G., Kongshavn, P.: The effect of dietary amino acid on the growth of tumors. Experientia, 37: 271-272, 1981.

33. Bounous, G., Batist, G., Gold, P.: Immunoenhancing property of dietary whey protein in mice: Role of glutathione. *Clinical and Investigative Medicine,* Vol. 12, 154-161, 1989.

34. Anderson, M.E.: Tissue glutathione. In CRC Handbook of Methods for Oxygen Radical Research. Boca Raton, Critical

Review: 317-29, 1985.

35. Meister, A., Anderson, M.E.: Glutathione, *Ann Rev Biochem* 52: 711-60, 1983.

36. Meister, A.: The antioxidant effects of glutathione and ascorbic acid. in Oxidative stress, cell activation and viral infection. Edit Pasquier C. et al., 101-111, 1994. Birkauser Verlag, Basel.

37. Bray TM, Taylor CO. Enhancement of tissue glutathione for antioxidant and immune functions in malnutrition. *Biochem Pharmacol* 47: 2113-23, 1994.

38. Puri RN, Meister A. Transport of glutathione, as γ-glutamylcysteinylglycyl ester, into liver and kidney. *Proc Natl Acad Sci USA* 80: 5258-60, 1983.

39. Anderson ME, Powric F, Puri RN, Meister A. Glutathione monoethyl ester: Preparation, uptake by tissues, and conversion to glutathione. *Arch Biochem Biophys* 239: 538-48, 1985.

40. Birnbaum SM, Winitz M, Greenstein JP. Quantitative nutritional studies with water-soluble, chemically defined diets. III. Individual amino acids as sources of "non-essential" nitrogen. *Arch Biochem Biophys* 72: 428-36, 1957.

41. Bounous, G., Gold, P.: The biological activity of undenatured whey protein: role of glutathione. *Clin. Inves. Med.* 14: 296-309, 1991.

42. Williamson JM, Boettcher B, Meister A. Intracellular cysteine delivery system that protects against toxicity by promoting glutathione synthesis. *Proc Natl Acad Sci USA* 79: 6246-9, 1982.

43. Droge W. et al. Role of cysteine and glutathione in HIV infec-

tion and cancer cachexia: therapeutic intervention with N-acetylcysteine. *Adv Pharm* 38: 581-600, 1997.

44. Mant TGK, Tempowski JH, Volans GN, Talbot JCC. Adverse reactions to acetylcysteine and effects of overdose. *Br Med J* 289: 217-19, 1984.

45. Eigel WN, Butler JE, Ernstrom CA, Farrell HM et al. Nomenclature of proteins of cow's milk. Fifth revision. *J Dairy Sci* 67: 1599-631, 1984.

46. Goodman RE, Schanbacher FL. Bovine lactoferrin in RNA: Sequence, analysis and expression in the mammary gland. *Biochem Biophys Res Commun* 180: 75-84, 1991.

47. Lomaestro BM., Malone M., Glutathione in health and disease: pharmacotherapeutic issues. *Ann Pharmacotherapy* 29: 1263-73, 1995.

48. Hirai R, Nakai S, Kikuishi H, Kawai K. *Evaluation of the Immunological Enhancement Activities of Immunocal.* Otsuka Pharmaceutical Co. Cellular Technology Institute, Dec. 13, 1990.

49. Baruchel S., Wainberg M.A., The role of oxidative stress in disease progression in individuals infected by the human immunodeficiency virus. *J. Leucocyte Biol.* 52: 111-114, 1992.

50. Halliwell B., Cross C.E., Reactive oxygen species, antioxidants and AIDS: Sense or speculation? *Arch. Intern Med.* 151: 29-31, 1991

51. Pantaleo G., Graciozi C., Fauci A.S., The immunopathogenesis of HIV infection. *NEJM* 338: 327-335, 1993.

52. Buhl R. et al. Systemic glutathione deficiency in symptom-free HIV positive individuals. *Lancet* 2: 1294, 1989.

53. Bounous G, Baruchel S., Falutz J, Gold P., Whey proteins as a food supplement in HIV-seropositive individuals. *Clin Invest Med* 16: 3, 204-209, 1993.

54. Baruchel S., Bounous G, Gold P.: Place for an antioxidant therapy in HIV infection, in Oxidative stress, cell activation and viral infection, Pasquier et al. (eds.) 311-321, 1994. Publ. Birkauser Verlag Basel

55. Baruchel S., Viau G., Olivier R., Bounous G., Wainberg M.A.: Nutriceutical modulation of glutathione with a humanized native milk serum isolate. Immunocal™: application in AIDS and cancer, in Oxidative Stress in Cancer, AIDS and Neurodegenerative Diseases, Ed. Montagnier L., Olivier R., Pasquier C., Publ. Dekker M. Inc., New York 447-461, 1998.

56. Herzenberg C.A. et al.: Glutathione deficiency is associated with impaired survival in HIV diseases. *Proc. Natl. Acad. Sci. USA* 96: 1967-72, 1997.

57. Bounous G.: Immunoenhancing properties of undenatured milk serum protein isolate in HIV patients. Presented at the 1997 International Whey Conference., Chicago, IL.

58. Flores S., McCord I.: In press in "Oxyradicals in Medical Biology, Ed. McCord I., Publ. J.A.I., Greenwich, CT.

59. Watanabe A, Higuchi K, Yasumura S, Shimizu Y et al.: Nutritional modulation of glutathione level and cellular immunity in chronic hepatitis B and C. *Hepatology* 24: 1883, 1996.

60. Richman P., Meister A.: Regulation of gamma-glutamylcysteine synthetase by nonallosteric inhibition by glutathione. In *Journ Biol Chem* 250: 4, 1422-26, 1975

61. Papenburg R., Bounous G., Fleiszer P, Gold P.: "Dietary milk

proteins whilst the development of dimethylhydrazine induced malignancy. *Tumor Biol.*, 11: 129-136, 1990

62. Russo A., Degraff W., Friedman N., Mitchell F.B.: Selective modulation of glutathione levels in human normal versus tumor cells and subsequent differential response to chemotherapy drugs. *Cancer Res.* 26: 2845-48, 1986.

63. Baruchel S., Viun G.: *Invitro* selective modulation of cellular glutathione by a humanized native milk protein isolate in mammal cells and rat mammary carcinoma model. *Anticancer Res* 15: 1095-1100, 1996.

64. Watanabe A., Treatment of chronic hepatitis using whey protein (non-heated). Presented at 16th Intl. Congress on Nutrition, Montreal, 1997.

65. Kennedy R.S., et al.: The use of whey protein concentrate in the treatment of patients with metastatic carcinoma: a Phase I – II Clinical Study. *Anticancer Res.* 15: 2643-50, 1995.

66. Hercbergs A., Brok-Simonif., Holtzmanf. Erythrocyte glutathione and tumor response to chemotherapy. *Lancet* 339: 1074-76, 1992.

67. Duncan B, Ey J, Holberg CJ, Wright AL et al.: Exclusive breastfeeding for at least 4 months protects against otitis media. *Paediatrics* 91: 867-72, 1993.

68. Frank AL, Taber LN, Glezen WP, Kasel GL et al.: Breastfeeding and respiratory virus infection. *Paediatrics* 70: 239-45, 1982.

69. Mather G, Gupta N, Mathur S, Gupta U. et al.: Breastfeeding and childhood cancer. *Indian Paediatrics* 30: 652-7, 1993.

70. Davis MK, Savitz DA, Graubard BI.: Infant feeding and childhood cancer. *Lancet* 1:365-8, 1988.

71. Bounous, G., Kongshavn, P.A.L., Taveroff, A., Gold, P.: Evo lutionary traits in human milk proteins. *Medical Hypothesis*, 27: 133-140, 1988.

72. McCamish MA., Bounous G., Geraghty M.E.: History of Enteral Feeding: Past and present perspectives. In Enteral and Tube Feeding Eds. Rombeau J.L. and Rolandelli R.H., Publ Saunders 1-11, 1997.

73. Lands L.C., Grey V.L., Smountas A.A., The Effect of Supplementation with a cysteine donor on muscular performance. Presented at the Association Pulmonaire du Québec, Quebec City, October 1998.

INDEX

A DOOR TO THE FUTURE

Dr. Bounous' *breakthrough* in cell-defense has opened a door for others to enter. More research will undoubtedly reveal the increasing value of this approach to nutritional pharmacology. In the meantime, we can benefit from the *real* glutathione revolution.

Ordering Information:

Breakthrough In Cell-Defense

Single copies......................$ 14.95US / $16.95CDN
10 or more copies:............. 10% discount

Shipping & handling: $3.00 per single copy

VISA/Mastercard orders:
24-hour Voice Mail
1-800-501-8516

Or mail your order to:

GOLDENeight Publishers
2778 Cumberland Rd, Unit: 206
Smyrna, GA 30080

Please allow six weeks for delivery.